Callan Cohen & Claire

ESSENTIAL
BIRDING
WESTERN SOUTH AFRICA

To the memory of Julie te Groen (1922–2000), whose selfless dedication to the Cape Bird Club encouraged a generation of new birders.

Struik Publishers (Pty) Ltd
(A member of the Struik New Holland Publishing Group)
Cornelis Struik House
80 McKenzie Street
Cape Town, 8001
Reg. No. 1954/000965/07

1 3 5 7 9 10 8 6 4 2

First published 2000

Project manager: Pippa Parker
Managing editor: Simon Pooley
Editor: Peter Joyce
Editorial Assistants: Giséle Raad, Sally Woudberg
Designer: Dominic Robson
Cartographers: Elaine Fick, Dominic Robson,
Callan Cohen and Claire Spottiswoode
Picture researcher: Carmen Watts

Reproduction by Hirt & Carter (Pty) Ltd, Cape Town
Printed and bound by CTP Book Printers, Cape Town

ISBN: 1 86872 524 3

Authors' Acknowledgements:

We are most grateful for years of encouragement and shared birding expertise from numerous friends,
especially John Graham, Marc and Diane Herremans, Jonathan and Sherran Rossouw, Peter Ryan, Peter
Steyn and David Winter, as well as David Allan, Tim Boucher, Jeff Cohen, Gruff Dodd, Morné du Plessis,
Mike Fraser, Phil Hockey, Jan Hofmeyr, Rob Leslie, Geoff Lockwood, Kirsten Louw, Liz McMahon, Sue
Maré, Giselle Murison, Ian Sinclair, Mel Tripp, Les Underhill, Phil Whittington, and all at the Cape Bird
Club and Percy FitzPatrick Institute of African Ornithology. We are also grateful to our parents, Mark and
Alice Cohen, and Christopher and Cécile Spottiswoode, for their support and encouragement of our early
interest. We are extremely grateful to Barrie Rose for generously contributing the pelagic table on p.35,
and, for their years of pelagic birding assistance and observations, we would like to thank Bruce Dyer,
Anne Gray and Trevor Hardaker. Numerous people gave helpful input to certain accounts and comment-
ed on the text: many thanks to Kirsten Louw, Mark Anderson, Mark and Alice Cohen, Paul Funston,
Sharon Hampson (Roberts' VII, proposed new bird names), Barrie Rose, Peter Ryan, Bernard Swart, Kim
Spottiswoode, Ross Wanless, David Wilcove and Sally Woudberg. We are also very grateful to Marilyn
Kennedy-McGregor and Edward Burn for their generous help, and would like to thank the Cape Bird
Club and BirdLife South Africa for endorsing the book. Finally, our thanks to all at Struik for their
tremendous help, and especially to Pippa Parker, Simon Pooley, Dominic Robson and Giséle Raad.

Contents

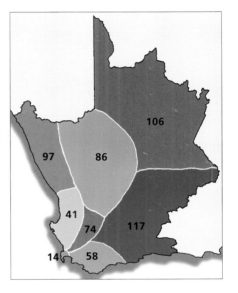

Birding in the Cape

Western South Africa is an extraordinarily biodiverse region, and an indispensable destination for the ecotourist in southern Africa. Indeed, there is no other area in Africa that offers such a high level of endemism in such an accessible setting. The region is well known among international and local birders alike for the remarkable suite of birds that is found here and nowhere else in the world: a staggering 42 of South Africa's 53 endemic bird species occur here, as well as 76 per cent of southern Africa's 181. Complimenting this unique birdlife is the scenic and botanical wealth of western South Africa, which is now acknowledged to contain two of the African continent's four biodiversity hotspots as defined by Conservation International.

In addition, local and international bird-watchers are inevitably drawn to the region by the tourism gem of Cape Town, and by the region's scenic and cultural diversity, well-developed infrastructure, high standard of accommodation, and excellent network of national parks and provincial and private nature reserves. A total of 613 bird species have been recorded in this region, and a two-week trip could be expected to yield in excess of 350 species. Indeed, over 220 species have even been recorded around Cape Town in a single day! Although the sheer diversity of southern Africa's more tropical eastern region is inevitably higher, most of the species found here have wide distributions and extend over much of Africa. The west, by contrast, is rich in species largely restricted to this region, making western South

The distinctive icon of Table Mountain welcomes visitors to the Cape.

African an essential destination in both global and local terms. Furthermore, this region offers representatives from seven of Africa's ten endemic and near-endemic bird families: Ostrich, Hamerkop, Guineafowl, Secretarybird, Mousebirds, Turacos, Woodhoopoes, African Barbets and Sugarbirds.

ABOUT THIS BOOK

Essential Birding was born out of frequent requests by both local and visiting birders for assistance: what are the best areas to visit, where are the best sites for the endemic birds, and how can we see them during a short visit to the region? We have done our best to answer these questions.

The region covered by this book combines the Western and Northern Cape provinces of South Africa, loosely referred to as the Cape. Our purpose is to provide the best sites for the region's characteristic birds and to link these together into practical routes suitable for a short visit to the region. Each route forms a chapter (see map on contents page), beginning with a regional map and introduction, and followed by site accounts accompanied by individual maps. At the beginning of each chapter is a short list of its most sought-after species ('Top Birds'), and the chapter is completed by a more detailed discussion of 'Select Specials'. In the latter, we have aimed to give the reader a feel for the best places to visit and techniques to employ in searching for these species (asterisked* page numbers alongside bird names in the text refer to these 'Select Specials' pages).

The annotated bird list at the end of the book lists all the species that have been recorded in the region, together with their alternative and scientific names, references to text citations, and status along each of this book's nine routes. Western South Africa forms part of the well-defined birding region of southern Africa, a region that includes South Africa, Namibia, Zimbabwe, Botswana, Lesotho, Swaziland and southern Mozambique (as covered by local field

Biomes of Western South Africa

See Overleaf

Savanna

Fynbos

Succulent Karoo

Nama Karoo

Afromontane Forest

Fynbos

Strandveld

guides). Throughout this book, we refer to 'endemics' as those birds restricted to greater southern Africa, unless explicitly stated otherwise. Near-endemics are those birds whose ranges extend only marginally beyond the borders of this subregion.

CHANGING BIRD NAMES

Southern Africa has long had distinctive indigenous names (often derived from Afrikaans) for all its birds, and many of these have filtered into official English usage. These include 'korhaan' for small bustards, 'dikkop' ('thick-head' or 'idiot') for thick-knees, and 'lourie' for turacos. Others have become internationally accepted, such as 'Hamerkop' ('hammerhead'). However, in the interests of global consistency, most southern African name peculiarities are set to conform to international nomenclature within the next few years. We list the imminent new names in the annotated checklists on pp.126–133.

CAPE BIRDING HABITATS

Five major biomes (broad-scale vegetation categories) occur in western South Africa: Fynbos, Succulent Karoo, Nama Karoo, Forest and Savanna. Within these biomes, the variation in vegetation, topography and human alteration has created a complex array of birding habitats. We have simplified this variation into five major birding 'habitats' that encapsulate the region's characteristic and endemic birds. Below, we discuss these categories and highlight those endemics (see p.5) that are linked to a particular habitat type.

1. Cape Floral Kingdom and Fynbos

The smallest of the world's six floral kingdoms, the tiny Cape Floral Kingdom is one of the richest biodiversity hotspots on earth, and is almost totally restricted to the area covered by this book. Despite occupying less than 0.05 per cent of the earth's land surface, this small pocket of diverse vegetation lying at Africa's southern extremity holds an astronomical 8 700 plant species. The winter-rainfall Cape Floral Kingdom encompasses all the vegetation types in the geographical area stretching from the Nieuwoudtville escarpment in the northwest through to Port Elizabeth in the southeast, and from the arc of mountain ranges to the sea. However, it is the shrublands of the Fynbos Biome, the largest and most prominent subset of the Cape Floral Kingdom, that lends this region its essential character.

Fynbos (Afrikaans for 'fine bush' – probably referring to the plants' spindly stems that are unsuitable for timber) is remarkable not only for its variety of plant species but also for its ecological peculiarities, among them its absolute dependence on regular fires, essential for successful germination of many fynbos plants, and the fascinating partnerships developed between plants and animals. Flower pollinators range from sugarbirds and bizarre long-tongued flies to nocturnal mice, while seed dispersal is facilitated by pheromone-laced seed coat-

Fynbos consists largely of the broad-leafed proteas (left), the reed-like restios (front, right) and the fine-leafed, heath-like ericas (along bottom).

ings that entice ants to plant them in their underground nests, ready to germinate following the next fire.

Although pristine areas of fynbos have a very low diversity of birds, there are some notable endemics to be seen, namely **Hottentot Buttonquail** (see p.23), **Cape Rockjumper** (p.73*), **Victorin's Warbler** (p.73*), **Cape Sugarbird** (p.33*), **Orange-breasted Sunbird** (p.33*), **Cape Siskin** (p.33*), and **Protea Canary** (p.57*). While not restricted to fynbos, the ranges of **Cape Francolin**, **Agulhas Long-billed Lark** (p.73*) and **Cape Bulbul** are largely centred on the Cape Floral Kingdom.

Fynbos vegetation is composed largely of three conspicuous plant groups: the large, broad-leafed proteas (favoured by **Cape Sugarbird** and **Protea Canary**); the low, small-leafed ericas (heath; favoured by **Orange-breasted Sunbird**, see picture on p.33); and the grass-like, clumped restios (brownish 'thatching reeds'). All are readily recognizable. For further information, visit the Kirstenbosch National Botanical Garden.

6

The true fynbos endemics are restricted primarily to the widespread mountain fynbos that occurs on rocky slopes and mountains of the Cape. Much rarer is lowland fynbos, which occurs on the flats of the coastal forelands. Here, too, Renosterveld occurs (p.63), a related vegetation type that has many bird species in common with fynbos, such as **Agulhas Long-billed** and **Clapper Larks**. The broad-leafed 'strandveld' ('beach vegetation') thicket, which occurs in a narrow band along the coast, is not related to fynbos. However, on parts of the West Coast it forms a mosaic with lowland fynbos, and many bird species are shared (p.43).

2. Succulent and Nama Karoo

The Karoo is a vast semidesert area that is divided into two botanically very different regions and dominates the arid western half of South Africa. It forms part of the most ancient desert system in the world, and is an open area of stony plains, scattered with small shrubs, punctuated by low dunes and hills ('koppies'), and is very sparsely inhabited. The Succulent Karoo Biome (*Tanqua Karoo* and *Namaqualand*) is characterized by small succulent plants, supported by low but predictable winter rainfall, whereas the summer-rainfall Nama Karoo Biome (*Bushmanland*, the southern *Kalahari*, and the Karoo

National Park in *Garden Route and Interior*) is dominated by grasses and low, woody shrubs. The Succulent Karoo Biome is one of Africa's biodiversity hotspots, and has the highest diversity of succulent plant species in the world.

Despite these fundamental climatic and vegetation differences, most Karoo bird specials occur in both biomes. Karoo endemics and near-endemics include **Karoo Korhaan, Ludwig's Bustard** (p.105*), **Red** (p.96*), **Barlow's** (p.100), **Karoo Long-billed** (p.13) and **Sclater's Larks** (p.96*), **Black-eared Finchlark** (p.96*), **Karoo** and **Tractrac Chats, Karoo Eremomela** (p.85*), **Cinnamon-breasted Warbler** (p.85*), **Namaqua Warbler** (p.85*), **Pale-winged Starling** and **Black-headed Canary** (p.105*). Numerous other species, such as **Karoo Lark** and **Rufous-eared Warbler**, are characteristic of this region, but also extend peripherally into other biomes.

3. Afromontane Forest

This biome is scattered discontinuously across central and east Africa's montane peaks, with the temperate forests of the Cape constituting its southern fragments. In the west of our region (*Cape Peninsula* and *Overberg*), pockets of afromontane forest survive in moist, fire-protected areas, and along the *Garden Route* large tracts extend along the coastal plain. Although these relict forest patches are rather species poor compared to those further north and east in Africa, **Knysna Warbler** (p.32*) and **Knysna Woodpecker** (p.72*) are almost restricted to this region.

More widespread Southern African forest endemics occurring here are **Forest Buzzard, Knysna Lourie** (p.125), **Chorister Robin** (p.125*), **Cape Batis,**

Succulents on a Tanqua Karoo koppie.

Emerald
Cuckoo

Arid savanna near Kimberley.

Olive Bush Shrike and Forest Canary, and other special birds include the Crowned Eagle, Narina Trogon (p.125*) and the dazzling Emerald Cuckoo.

4. Arid Savanna

Perhaps Africa's most characteristic vegetation type, this biome forms an intermediate between grassland and woodland, and occupies the northeast of our region, the Kalahari. Rainfall is in the form of summer thunderstorms that support good grass cover below a varying density of thorn-trees, perhaps most characteristically the camel-thorn (*Acacia erioloba*). Although savanna supports a diverse bird community, most of these species are widely distributed in southern Africa and further afield. However, the savanna in this region is characteristically arid, and endemics include Short-clawed Lark, Kalahari Robin, Ashy Tit, Marico Flycatcher and Crimson-breasted Shrike. The species composition present at a given locality depends on the relative density of grass and trees; consult *Kalahari* for details.

5. Coastal and Wetland

Both the Atlantic and Indian oceans flank this region, merging at Africa's southern-most point, Cape Agulhas (p.59). The productive Benguela Current surges up the Atlantic coast, bringing chilly, nutrient-rich waters from Antarctica, while the warmer Agulhas Current moves down the east coast of Africa from more tropical climes. The birds endemic or near-endemic to the plentiful waters of the Benguela Current of southern Africa's west coast are African Penguin (p.32*), Cape Gannet, Cape, Bank (p.21*) and Crowned Cormorants, African Black Oystercatcher (p.33*), Hartlaub's Gull and Damara Tern; consult the *Cape Peninsula*, *West Coast*, *Overberg* and *Namaqualand* chapters. Furthermore, huge numbers of migrant pelagic seabirds are attracted to offshore waters (see *Seabirding*).

UNDERSTANDING PLACE NAMES	
Afrikaans is the most widely spoken language in western South Africa and this list will help in translating many rural place names.	
berg:	mountain
burg:	town
bos:	forest/bush
dorp:	village
fontein:	fountain, spring
kloof:	gorge
kop; (koppie)	peak, hill; (hillock)
olifant:	elephant
poort:	'gateway' through mountains
strand:	beach
tier/tyger/luiperd:	leopard
veld:	natural countryside
vlei:	wetland, often seasonal

Wetland habitats hold two endemic waterfowl species.

PLANNING YOUR TRIP

BIRDING SEASONS
The best time to go birding in the Western Cape is undoubtedly springtime. This is because the majority of the region receives its rain in winter, and the animals and plants must use the small window of opportunity to breed while temperatures are sufficiently high and moisture is still in good supply. Birding picks up significantly towards the end of August, and the very best birding months are September and October. However, the weather at this time of year is unpredictable, as the winter rains often linger. Spring is also the time to witness the West Coast and Namaqualand flower displays, with the peak flower season varying from late August to late September (p.99).

Good birding continues into the summer, although things slowly become less active when water supplies dwindle as this hot, dry season progresses. In contrast, Bushmanland, the Nama Karoo and the Kalahari experience summer thundershowers, resuscitating the grasses and revitalizing the birdlife. Autumn is thus the best birding season in these regions, although spring is also very productive. As autumn progresses into winter, pelagic birding off the Cape becomes increasingly exciting (see the monthly table on p.35), but the persistent, rainy cold fronts that buffet the coastal area in midwinter can make birding impossible for days at a time.

ITINERARIES
Foreign birders visiting the western portion of South Africa as part of a longer tour around the country will need a very bare minimum of five days based in Cape Town. Those interested in exploring *Namaqualand*, *Bushmanland* or the *Kalahari* will clearly need more time. Good loops through these areas would ideally need at least three days each, plus additional driving time if travelling to the Kalahari from Cape Town.

Those in Cape Town on business and with only a few hours to spare can still see a good selection of Cape specials at sites close to the city. The Kirstenbosch National Botanical Garden (p.15) is undoubtedly the best such locality: it is a mere 15 minutes' drive from the city centre. Furthermore, it is easily accessible by public transport (consult Cape Town Tourism, p.136). Other good localities within 30 minutes of the city are the Strandfontein sewage works (p.26), Kommetjie (p.21), Boulders Beach (p.24) and even the Atlantic shoreline right outside the city centre (p.31) – within walking distance of the Waterfront. For those with three or more days to spend in Cape Town, we would recommend a day each on the *Cape Peninsula*, the *Tanqua Karoo*, and either the *West Coast* or the *Overberg* region. A very desirable addition would be a day out at sea (*Seabirding*).

While an ideal visit to western South Africa would encompass most if not all of the routes we describe in this book, this is clearly

Karoo scene

not practical for those with limited time. Here (right) we propose two itineraries for the serious birder aiming to maximize coverage of endemic species.

HEALTH, SAFETY AND TRAVEL

Please consult local tourist information offices (see p.136) for further information on the topics discussed below. None of the sites covered in this book are unusually dangerous, although we do urge visitors to be cautious and alert, particularly in Cape Town and on the Cape Peninsula, as birders' inevitable need to carry conspicuously valuable equipment makes them potential targets for casual muggings. Sites that are perhaps best not visited alone are Strandfontein sewage works (p.26) and Sir Lowry's Pass (p.60), although years of birding activities at these sites have yet to result in any incidents. Petty theft is common; never leave bags or birding equipment unattended on car seats.

Visitors may be relieved to hear that there is no malaria in the region covered by this book, although it has been recorded previously in the Kalahari Gemsbok National Park (p.107; enquire before visiting). Should you be lucky enough to see a snake or scorpion, please be cautious as some species are potentially dangerous.

Road infrastructure in South Africa is excellent, and none of the routes recommended in this book require a four-wheel-drive vehicle. However, there might be potential driving hazards in rural areas for those unaccustomed to unsurfaced roads. We refer the reader to

ONE-WEEK ITINERARY
Day 1: *Cape Peninsula*: Kirstenbosch, Cape of Good Hope, Boulders
Day 2: *Seabirding* boat trip
Day 3: *West Coast*
Day 4: *Cape Peninsula* in morning: Constantia Greenbelt. Then to Ceres via Paarl and Bain's Kloof.
Day 5: Ceres – *Tanqua Karoo* – Cape Town
Day 6: *Overberg*: Sir Lowry's Pass, Overberg farmland loops to De Hoop, Swellendam
Day 7: *Overberg*: Grootvadersbosch, return to Cape Town

TWO-WEEK ITINERARY
Day 1: Kirstenbosch, Cape of Good Hope, Boulders (*Cape Peninsula*)
Day 2: *Seabirding* boat trip
Day 3: Sir Lowry's Pass – De Hoop – Swellendam (*Overberg and South Coast*)
Day 4: Grootvadersbosch – Cape Town (*Overberg and South Coast*)
Day 5: Cape Town – Darling and West Coast National Park
Day 6: West Coast – Kransvlei Poort– Brandvlei (*West Coast and Bushmanland*)
Day 7: Brandvlei – Kenhardt (*Bushmanland*)
Day 8: Kenhardt – Kalahari Gemsbok National Park (*Bushmanland and Kalahari*)
Day 9: Kalahari Gemsbok National Park
Day 10: Kalahari Gemsbok National Park
Day 11: Kalahari Gemsbok National Park – Pofadder (*Kalahari and Bushmanland*)
Day 12: Pofadder – Port Nolloth – Springbok (*Namaqualand*)
Day 13: Springbok – Kamieskroon
Day 14: Kamieskroon – Cape Town

If you wish to visit the *Garden Route* forests, it is equally possible to continue on to them from Swellendam. One can then link to the central *Bushmanland* region via Swartberg Pass and the Karoo National Park, rather than from the *West Coast* as above, and instead visit the *West Coast* in a day trip from Cape Town.

pp.77 and 87. If you need to travel long distances in the Northern Cape, please be sure to carry a good water supply in case of breakdowns. In the latter respect, be aware that mobile phone coverage in remoter areas is far from complete.

A wide range of accommodation is available across most of the region, especially in the coastal areas; consult tourist information for further details (p.136). Accommodation in the national parks and most nature reserves is superb and great value for money. All small rural towns have service stations and basic, inexpensive hotels, and many have municipal campsites. Tap water in towns is invariably potable. Public transport is limited, often unsafe, and best avoided.

BIRDING ETIQUETTE
With the ever-increasing popularity of birding in South Africa, there is inevitably pressure on certain well-visited sites. While the roadside often provides excellent birding in rural areas, please ask permission at the nearest farmhouse if you would like to enter private land. Playback of bird calls is a very helpful birding tool, and limited tape playing is unlikely to have a detrimental effect in most cases. However, we urge birders to refrain from excessive playback, especially during the breeding season and at popular sites such as those for Knysna Warblers on the Cape Peninsula (p.19), and Cinnamon-breasted Warblers at Katbakkies (p.79).

Mural with a message: watch out on beaches for breeding African Black Oystercatchers.

INFORMATION
Lists of recommended field guides, sources for further reading and useful contacts appear on pp.135 and 136.

GETTING INVOLVED

BIRD CLUBS
Local birders are encouraged to join the Cape Bird Club, currently Africa's largest bird club, and one of 20 regional branches of BirdLife South Africa (see Useful Contacts, p.136). This is a friendly and informal club; regular club activities include a monthly evening meeting in Newlands (Cape Town) and numerous monthly half-day outings and other events. All members receive a quarterly magazine. The Cape Bird Club and BirdLife South Africa are committed to the conservation of birds and their habitats.

INTERNET BIRDING
The local e-mail list, *Cape BirdNet*, provides an active forum of over 100 birders for local observations, rarity updates, trip reports and local birding events in the region covered by this book (to find out more, and to join, visit **www.birding-africa.com**, which also supplies a comprehensive list of southern African birding websites). E-mail lists provide the visitor with all sorts of helpful titbits for trip preparation. You may also wish to join *SA BirdNet*, a similar birding forum serving the whole of southern Africa (to join, contact norman@nu.ac.za).

REPORTING SIGHTINGS
Please report all sightings of colour-ringed birds, and ring numbers from dead birds, to SAFRING (see p.136). Birders can also contribute to other valuable projects run by the Avian Demography Unit. If you see any species that are either very rare or not recorded for this region, please contact the local or national rare bird committees via the Cape Bird Club and BirdLife South Africa respectively (p.136).

TAXONOMY FOR BIRDERS
To tick or not to tick?

Our perception of the birds of the Cape and the wider southern African region is currently fraught with controversy and confusion. Not only are many of the common names undergoing change (p.5), but there is the added complication of 'new' species: **Barlow's Lark** (p.100), **Southern Black Korhaan** (p.57*) and **Agulhas Long-billed Lark** (p.73*) are just a few of the nearly 20 'new' birds that have emerged in field guides over the past few years. These should not be confused with totally undescribed forms, such as the **Long-tailed Pipit** (p.115) that was recently discovered near Kimberley. Rather, these 'new' species represent subspecies that have been 'split' and elevated to full species status.

The 'species', a unit integral to science, conservation and indeed birding, may not be as absolute as we would imagine it should be. The way scientists view species ('species concepts') has always been subject to much debate, and whether similar forms should be designated as species or subspecies is in practice frequently a matter of opinion.

The once dominant and undoubtedly valuable Biological Species Concept places emphasis on the ability of individuals to interbreed: forms that can interbreed and produce fertile offspring are regarded as members of the same species. One weakness of this concept is that it fails to deal effectively with isolated populations that would never naturally come into contact and interbreed. It also fails to deal objectively with natural hybridization, and strict supporters advocate that hybridizing groups must belong to the same species. It is thus often associated with 'lumping', which is the downgrading of similar species into subspecies of a single species. A local example of a situation where application of the Biological Species Concept might be misleading is that of the recently discovered narrow hybrid zone between **Karoo Lark** and **Barlow's Lark** (p.100). These two distinctly different species are not even each other's closest relatives, and their inappropriate 'lumping' into one species would obscure the myriad fascinating differences shown between these forms.

Among ornithologists, the relatively recent Phylogenetic Species Concept is gaining popularity. It places emphasis on consistent differences (even if small) between forms, and is applied within the framework of a phylogeny (a family tree of evolutionary relationships). It is based on the indirect argument that if forms show distinct differences, then they must be separate, because mixing results in shared features, not distinct ones. Because the Phylogenetic Species Concept recognizes small differences between forms, it is often linked with 'splitting', which is the process of elevating subspecies to full species status. Both species concepts have their drawbacks and are currently causing much debate among scientists.

Changing species concepts are only part of the reason for the recent appearance of 'new' species. Thorough new research spearheaded by ornithologists at the Percy FitzPatrick Institute of African Ornithology at the University of Cape Town, Dr Peter Ryan and Professor Tim Crowe, and molecular geneticist Dr Paulette Bloomer, is revealing previously undetected differences among closely similar groups of southern African birds (see Further Reading, p.135). Central to these studies are genetic techniques that probe deep into the birds' evolutionary histories to obtain estimates of their uniqueness, as part of a multifaceted approach that also includes evidence from plumage, size, calls and behaviour, resulting in taxonomic decisions that are likely to stand the test of time.

Relatively recent changes include the resurrection of a historical split between **Northern (White-quilled)** and **Southern Black Korhaans** (p.57*), and the split of **Knysna Lourie** (p.125*) from a host of green turacos that occur further north in Africa. Thorough lark research has revealed **Barlow's Lark** (p.100) as a new species; by

'Namaqua' Olive Thrush

'Cape' Olive Thrush

contrast, the different forms of **Red Lark** (p.96*) have been shown to be sub-species. The endemic **Long-billed Lark** *Certhilauda curvirostris* has been demonstrated to comprise five species, of which the **Cape** (western coastal lowlands, pp.52, 101), **Agulhas** (Overberg, p.73*) and **Karoo** (central arid regions, pp.89, 125) **Long-billed Larks** occur in the Cape.

There is no doubt that there are differences in many of our bird species that warrant the further recognition of 'new' species, which only thorough investigation will be able to reveal. However, examination of previous classifications and striking variation in voice and plumage have allowed us to speculate on possible future splits of which visiting birders might wish to be aware. Below we use the scientific notation of *genus (species) subspecies*, with the species name bracketed in the case of disputed splits.

Firstly, it is likely that certain African subspecies of globally widespread birds may become split, such as **Kelp Gull** *Larus (dominicanus) vetula*. See *Seabirding* (p.37) for further examples. Likewise, some South African subspecies may be split from their counterparts further north in Africa, such as **Cape Eagle Owl** (p.105*).

However, it is local variation among species here in South Africa that will prove most interesting to birders in the region. Birds that have previously been separated as different species by some authors, but which are lumped in current southern African field guides, include **Bradfield's Lark** *Mirafra (sabota) naevia*, the large-billed group of sub-species of the **Sabota Lark** that occur in the drier western areas of the Northern Cape. Furthermore, **Clapper Lark** has been historically split into three species occurring in the Cape: **Cape Clapper Lark** *Mirafra apiata apiata* (West Coast, p.44), **Adendorff's Clapper Lark** *Mirafra (apiata) adendorffi* (Namaqualand, p.104) and **Highveld Clapper Lark** *Mirafra (apiata) hewitti* (Kalahari and Kimberley, p.116*). A further distinctive subspecies occurs in the Overberg (p.64), namely *Mirafra (apiata) marjoriae*.

South African **Olive Thrushes**, although currently lumped as one species, have previously been classified as two: **Cape Thrush** *Turdus (olivaceus) olivaceus* (Cape Town eastwards through Overberg and Garden Route), and **Namaqua Thrush** *Turdus (olivaceus) smithi* (parts of the West Coast, Namaqualand, and from the Kalahari to Kimberley). **Cape White-eye** has also been previously split, with two species occurring in the Cape: **Cape White-eye** *Zosterops (pallidus) capensis* (Cape Peninsula and West Coast to the Garden Route; shows grey flanks) and **Pale White-eye** *Zosterops pallidus pallidus* (Namaqualand and along the Orange River, to Kimberley; shows peach-orange flanks). Further discussion is provided in the text for **Black-rumped/Hottentot Buttonquail** (p.23), **Short-toed Rock Thrush** (p.124), **Cloud Cisticola** (p.57*), and **Black-headed Canary** (p.105*). A number of other species also show significant local variation and may yield interesting patterns; visit the 'African Bird Taxonomy' website for news of updates (**www.birding-africa.com**).

Cape Peninsula

'...the fairest Cape we saw in the whole circumference of the earth.' SIR FRANCIS DRAKE, 1580

Often considered one of the most scenic stretches of landscape in the world, the Cape Peninsula ranks among Africa's premier tourism destinations. For birders, it provides easy access to a good selection of coastal and mountain specials, and some of the world's best sites for such highly localized endemics as **Knysna Warbler, Hottentot Buttonquail** and **Cape Siskin,** and seabirds such as **African Penguin.**

A narrow, 75-km long strip of land separating the cold Atlantic upwelling from the waters of False Bay, the Peninsula's landscape is dominated by a rugged mountain chain, culminating at its northern end in the famously geometrical massif of Table Mountain. Cradled between this renowned landmark and its flanking peaks – Lion's Head and Devil's Peak – are Cape Town's city centre and harbour, site of the first colonial settlement in southern Africa and now the country's cultural and tourism epicentre.

Rising to 1 086 m and sculpted from delicately coloured sandstone, the Peninsula's mountains are clad in the extraordinarily diverse fynbos vegetation that is unique to the southern Cape region (see p.6). Table Mountain alone supports a

staggering 2 600 plant species, more than the entire British Isles. Despite residential development at lower altitudes, much pristine mountain landscape is protected in the newly proclaimed Cape Peninsula National

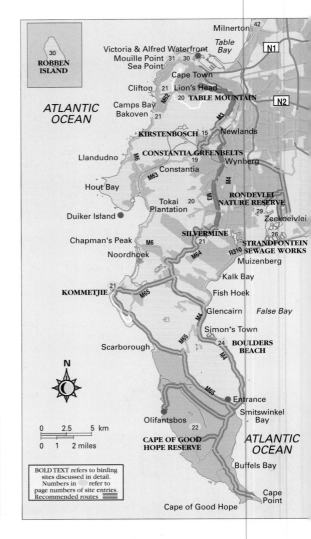

TOP BIRDS

African Penguin, Bank Cormorant, African Black Oystercatcher, Hottentot Buttonquail, Antarctic Tern, Knysna Warbler, Cape Sugarbird and Cape Siskin.

BOLD TEXT refers to birding sites discussed in detail. Numbers in refer to page numbers of site entries. Recommended routes

Above: Table Mountain remains one of the world's biodiversity treasures, despite South Africa's second largest city nestling at its foot. Below: Malachite Sunbird on a King Protea.

Park that runs, discontinuously, from Table Mountain to the Peninsula's tip at the Cape of Good Hope, and which is destined for recognition as a World Heritage Site. The coastline, spectacularly rugged in places, is punctuated by numerous idyllic beaches.

Dedicated birders with limited time can fit in an excellent day's birding on the Peninsula by starting early at the Kirstenbosch National Botanical Garden and Constantia Greenbelts before proceeding, via Kommetjie, to the Cape of Good Hope reserve for lunch, and finally winding up at Boulders Beach in the late afternoon. However, more relaxed visitors wishing to combine birding with general sightseeing could easily spread this programme over two or more days, expanding it to include the Table Mountain cableway, a boat trip to Robben Island or a visit to the city itself. The only site that, ideally, requires an early start is Kirstenbosch, as by mid-morning birds are less visible and tourists more so. A visit to the very productive Strandfontein sewage works is a must for those with an interest in waterbirds. Pelagic seabird trips (p.38) depart from the Cape Peninsula, either from Simon's Town or from Hout Bay. Although not

included in this chapter, Sir Lowry's Pass (p.60), Paarl (p.82) and Rietvlei (p.42) are also conveniently explored using Cape Town as a base, and may be combined with the above sites.

KIRSTENBOSCH NATIONAL BOTANICAL GARDEN

Widely recognized as one of the world's finest botanical gardens, Kirstenbosch would be an essential destination for its pleasing landscapes and spectacular floral displays alone. Additionally, the well-maintained gardens and adjacent fynbos and indigenous forest support an attractive diversity of bird species. Here, it is possible to approach a number of Cape endemics at close quarters, including such desirable species as **Cape Sugarbird** (p. 33*), **Orange-breasted Sunbird** (p. 33*) and **Cape Francolin.**

The modern Visitors' Centre (①) on site map overleaf), located near the main entrance, is well worth exploring. Here one can enjoy a meal, browse the well-stocked tourist shop, consult the information kiosk, or search the excellent bookshop – offering arguably Cape Town's best selection of natural history titles. The Visitors' Centre is also home to a world-class glasshouse (the Conservatory), which contains a variety of succulent plants from South Africa's arid areas. Ranging from a huge baobab to tiny stone-plants, this fascinating display is highly

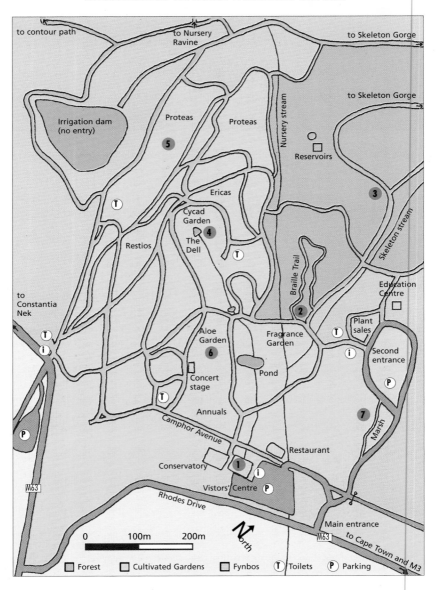

Forest ☐ Cultivated Gardens ☐ Fynbos ☐ (T) Toilets (P) Parking

recommended for those with even the most casual botanical interest. A pair of **Spotted Eagle Owls** nests annually in the camphor trees just behind the Visitors' Centre, and the staff usually know their precise whereabouts.

Castle Buttress stands sentinel above the gardens, flanked by Skeleton Gorge and Nursery Ravine. These are two of the numerous gorges that cut through the steep and moist eastern face of Table Mountain,

Above: Castle Buttress towers above Kirstenbosch, and is flanked by forested ravines where Cape Batis (below) occurs.

providing welcome refuge from the often baking summer heat of the surrounding fynbos. In the ravines, two metres of annual rainfall maintains pockets of pristine afromontane forest, through which acidic, fast-flowing streams course over moss-swathed boulders and decaying trunks festooned with ferns. Clinging, well-concealed, to the shaggy moss of these streams is the very rare Table Mountain Ghost Frog *(Heleophryne rosei)*, which is globally restricted to just a handful of gorges on these eastern slopes. Despite the idyllic setting, however, birders may find the Peninsula's forests disappointingly quiet relative to the diversity that abounds farther east – although there are rewards for those who are patient.

The most accessible forest is along the Braille Trail (② on map opposite) an easy walk that loops gently through the trees. It begins on a broad gravel path opposite the Fragrance Garden, which is just a short stroll across the lower lawns from the main entrance and Visitors' Centre. It is best to start here as early as possible, as human activity increases steadily during the morning. **Knysna Warbler** (p.32*) sometimes lurks along the streams and thickets here, but the Constantia Greenbelts (p.18) and Skeleton Gorge (p.19) are more

reliable places to search for it. **Cinnamon Dove** is a true forest species that is often found shuffling through the leaf litter in the dappled shade of the understorey. For the best chance of seeing it, make an early start up the gravel track mentioned above (②), concentrating your search at the sharp corner after about 300 m (③). Other forest birds found here and along the Braille Trail also extend down into the formal gardens. These include **Rameron Pigeon**, **Sombre Bulbul**, **Olive Thrush** (p.13), **Cape Batis**, **Paradise** and **Dusky Flycatchers**, and **Forest Canary** – the latter a relatively recent arrival on the Peninsula (see box, p.26). The whole of Kirstenbosch is also a fine area for raptors: early mornings and evenings are when forest hawks, such as **Red-breasted** and **Black Sparrowhawks** and **African Goshawk**, are most likely to be seen (see box, p.18).

Next, visit the small, tranquil forest patch, known as the Dell, which surrounds Colonel Bird's Bath, a clear spring welling up in the middle of the gardens (④). The huge trees overlooking the spring invariably hold perched **Rameron Pigeons**. From the Dell, one can ascend through the jumble of the Cycad Garden into the upper part of Kirstenbosch (⑤), where protea and erica plants abound. Their flowerheads are adorned with **Cape Sugarbirds** and **Orange-breasted Sunbirds** respectively, whose approachability will be much appreciated by photographers. Fynbos vegetation is globally renowned for its remarkable diversity of bird-pollinated plant species, and Kirstenbosch is a perfect place to observe this mutualism in action.

Make your way downhill from the protea section, along the right

Forest haunt of the Knysna Warbler.

PENINSULA RAPTORS

The forested slopes and rocky cliffs of the Cape Peninsula are prime raptor-watching areas. A number of **Black Eagle** pairs still nest on the mountains, and their distinctive silhouette can be seen gracing the skies anywhere along the Peninsula's rugged spine (pp. 19, 21). **Peregrine Falcon** (see picture below and pp. 21, 23) is unusually common here, sharing the skies with other cliff-nesting species including **Lanner Falcon**, **Rock Kestrel** and **Jackal Buzzard**. The two falcons commonly hunt in suburbia, where a flurry of scattering doves often betrays their presence.

The plantations and indigenous forests are the haunt of the agile **Red-breasted Sparrowhawk** and **African Goshawk**, both of which have adapted well to wooded suburbia. Although they are both common, the latter is more often seen due to its conspicuously noisy early-morning display flight. Ideally, familiarize yourself with the calls of both these species, and scan the skies often, especially in the early morning and evening (see pp 17 and 20).

Black Sparrowhawk is a recent arrival on the Peninsula and is still relatively scarce. See p. 20 for discussion on **Steppe**, **Forest** and **Honey Buzzards**.

hand edge of the Dell, following one of the myriad paths that eventually lead to the lower gardens. Common and conspicuous birds of the cultivated gardens are **Cape Francolin**, **Helmeted Guineafowl**, **Cape Robin**, **Karoo Prinia**, **Southern Boubou** and **Lesser Double-collared Sunbird.** Species found here that are characteristic of more wooded environments in the Cape are

Red-chested Cuckoo (calling September–November), **Black Saw-wing Swallow** (summer), **Speckled Mousebird** and **Bully Canary**. Passing the Aloe Garden (⑥), look carefully at Kirstenbosch's unusual **Fiscal Shrikes**, which show a patchy white eyebrow, characteristic of the northern Cape subspecies. Before leaving, it is worth visiting the small marsh (⑦) that lies below the old parking area. Here, the noisy **Grassbird**, **Levaillant's Cisticola** and **Common Waxbill** vie with the ceaseless chorus of Cape Chirping Frogs *(Arthroleptella lightfooti)* and Clicking Stream Frogs *(Strongylopus grayii)*.

The forests high above Kirstenbosch are the haunt of the localized and reclusive **Knysna Warbler** (p. 32*), a southern Cape endemic which lurks in the forest understorey, giving a beautiful song but obstinately resisting attempts at a clear view. Birders who enjoy hiking and decide to venture up to the top of Skeleton Gorge in search of **Knysna Warbler** might consider a round trip, returning behind Castle Buttress and down Nursery Ravine. This moderately strenuous half-day walk along orchid-lined streams quickly elevates one from the hustle and bustle of the city suburbs below and into the calm sanctum of Table Mountain's southern reaches. Although **Knysna Warbler** occurs along the length of the Skeleton stream, it is perhaps most easily seen in the bracken-dominated section at the head of the gorge, where the path climbs out of the rocky riverbed. On the mountain top, a pair of **Black Eagles** regularly hunts, **Black Swifts** nest in the cliffs at the top of Nursery Ravine, and **Cape Siskin** (p.33*) and **Ground Woodpecker** (p.105*) occur on the rocky crags of the buttress proper. The unrivalled elusiveness of the **Striped Flufftail** will become immediately apparent to anyone following up the bizarre cackles and hoots that emanate from the denser fynbos.

CONSTANTIA GREENBELTS

The forest patches which cloak much of the eastern slopes of Table Mountain extend down into the exclusive residential suburb of Constantia, south of Kirstenbosch (see regional map, p.14), where interlinked 'greenbelts' have been set aside to preserve the area's natural character. The forest and thickets along these greenbelts are the Peninsula's best sites to see the elusive **Knysna Warbler** (p.32*), **Buff-spotted Flufftail**, **Wood Owl**, and a host of other forest specials.

The largest section of the greenbelts is known as De Hel (① on site map, below). Park in the small area next to the 'Greenbelt' signboard and walk to the right on the broad, descending path which meanders down about 200 m to a small, vegetated stream at ②. The upper reaches of this stream can also be reached 100 m further along the path that leads from the left of the parking area ③. The secretive **Knysna Warbler** is found in the streamside thickets, where it creeps around low in the vegetation, often walking on the ground. This sombre skulker's strikingly beautiful song is the key to pinpointing its position, and tape playback often entices it closer. Excessive playback, however, can cause disturbance to the birds and we urge birders to act considerately – please don't use calls unnecessarily. Look out for **Cinnamon Dove** walking noisily on the forest floor in this area. Other birds seen regularly between here and the parking area include **Cape Siskin** (p.33*), **Forest Canary**, **Cape Batis**, **Sombre Bulbul**, **Dusky** and **Paradise Flycatchers**, **Rameron Pigeon**, **Redchested Cuckoo** (vocal from

Constantia Greenbelts

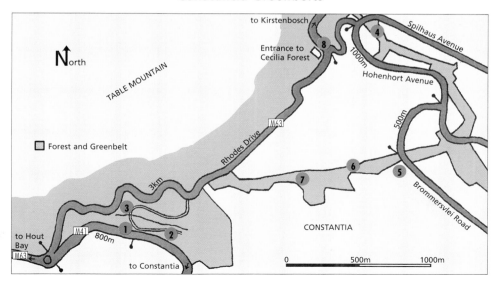

September to December), **Lesser Double-collared Sunbird** and the introduced **Chaffinch** (see box, p.31). This is also one of the few sites on the Peninsula where **Bar-throated Apalis** and **Swee Waxbill** occur, although they are scarce here.

Another good part of the greenbelts for **Knysna Warbler** is at ④ (again, park at the 'Greenbelt' signboard). A footpath leads down the densely vegetated stream along which the birds lurk.

Buff-spotted Flufftail, a relatively recent discovery on the Cape Peninsula, is another star bird of the Constantia Greenbelt. This is a legendary skulker, and is among Africa's hardest birds to see. Males are best searched for on summer nights when their ventriloquial, hooting call emanates from dense tangles of vegetation, often from a position up to 3 m above the ground. Park at the point where the greenbelt crosses Brommersvlei Road (⑤) and walk west for 300 m to a wooden bridge over a small wetland (⑥). In the evenings, a number of **Buff-spotted Flufftails** can be heard calling along a stretch from here up to the tar road that crosses the greenbelt (⑦). **Wood Owl**, **Spotted Eagle Owl** and the occasional **Fiery-necked Nightjar** can be heard at night.

Both **Red-breasted Sparrowhawk** and **African Goshawk** are common throughout this area. This is an excellent spot to get clear views of **African Sedge Warbler**, **Burchell's Coucal** (listen for its bubbling call) and **Common Waxbill**.

Cecilia Forest (main entrance at ⑧), a walking spot popular among Capetonians, consists predominantly of timber plantations with small patches of indigenous forest holding out along the streams. **Chaffinch** and **Cape Siskin** (p.33*) are particularly common in the plantations, and the open patches with a clear view of the skies are the best places to look overhead for soaring **Red-breasted Sparrowhawk** and the rare **Honey Buzzard**, interspersed among the much more common **Steppe Buzzard** and the occasional **Forest Buzzard**.

Raptorphiles may enjoy the identification challenge posed by the resident and migrant buzzards. **Forest Buzzard** is a rare resident of the Peninsula's forests, and is best seen at Tokai Plantation (take the Tokai off-ramp from the M3 and continue all the way to the mountain, turning left, at the Cape Dutch manor house T-junction, towards the Arboretum). The migrant **Steppe Buzzard** far outnumbers it during summer, and considerable skill is required to distinguish the two species. It is also always worth keeping an eye out for the scarce **Honey Buzzard**, whose presence in the Cape was revealed only in the mid-1980s, when it was discovered by John Graham at Tokai. Small numbers of this species can be found anywhere in forests on the Peninsula, although Tokai and Cecilia Forest are the places where they are seen most frequently.

TABLE MOUNTAIN

Cape Town's most famous symbol (since its days as a landmark to early European seafarers) is now easily accessible to visitors. You can ascend in a few minutes in an impeccably modern cable car, or make your way up the excellent network of hiking paths of every level of severity.

The wind-buffeted plateau can be explored by following the well-marked, surfaced paths that lead from the cableway station. The fynbos here holds a remarkably low density of birds, although **Orange-breasted Sunbird** (p.33*), **Neddicky** and **Grassbird** are usually reasonably common. **Red-winged Starlings, Rock Pigeons** and

Red-winged Starlings

BANK CORMORANT

This elegant Benguela endemic, which has suffered a massive population decline (only 4 900 breeding pairs remain), has a propensity for the unusual: it is the only cormorant to build its nest from fresh kelp (seaweed), which it plasters to seaside boulders with its own droppings. Moreover it is unique among birds in that its extraordinary turquoise eyes change to red from top to bottom as it matures, so that some individuals have bizarrely two-tone eyes!

Care needs to be taken in identifying this species, as the white-rumped breeding plumage is absent on most individuals and is not a good field character. It is, however, readily distinguished by its lack of any bare facial skin, pot-bellied appearance, and often-present small, bumpy crest. Reliable sites to see this species include Kommetjie (p. 22), Bakoven (p. 31), Stoney Point (p. 62), and on the West Coast (p. 46).

'dassies' (Rock Hyrax, *Procavia capensis*) compete for food scraps from tourists. **Rock Martin** and **African Black** and **Alpine Swifts** fly overhead. Visitors to the plateau should also keep alert for raptors: **Black Eagle**, **Peregrine Falcon** and **Rock Kestrel**, along with **White-necked Raven**, all nest on cliff faces in this vicinity and are regularly seen in flight.

Those with the time available may consider a walk up Lion's Head, an excellent site for **Cape Siskin** (p. 33*), **Ground Woodpecker** (p. 105*) and **Black Eagle**.

SILVERMINE

Travelling from Cape Town to Kommetjie or the Cape of Good Hope, one can travel via Ou Kaapse Weg ('Old Cape Road', the M64), a scenic pass that straddles the Peninsula's mountainous backbone and provides a convenient place to explore the mountains without having to walk too far. At the top there is a long, level stretch and a viewing point ('Silvermine' sign) before the road begins to curve and descend. Turn left along a short road that leads to a parking area. The adjacent rocky ridge is a fine site for **Ground Woodpecker** (p.105*), **Cape Siskin** (p.33*), **Familiar Chat** and **Peregrine Falcon**.

KOMMETJIE

Kommetjie is a small seaside village on a rocky promontory on the west coast of the Peninsula, much favoured by beach-walkers, horse-riders, anglers, surfers, and the few hardy swimmers who brave its usually icy Atlantic waters. To the north lies the pure white 4-km expanse of Noordhoek beach, also known as Long Beach (a great walk for those in a contemplative mood), and to the south a picturesque mountainside road leading around Slangkop peak to Scarborough and the Cape of Good Hope reserve. For birders, Kommetjie provides convenient access to a number of endemic or localized coastal species, notably **Bank Cormorant** and **Antarctic Tern** (winter).

Non-breeding Antarctic Tern: a winter visitor.

Entering Kommetjie from the east on the M65, turn right down Van Imhoff Road (at the sharp bend opposite the hotel). Continue to a prominent parking area on the left, where a path leads onto the rocky promontory. Stone Age people built rough rock fish-dams here; today, this jumble of lichen-splattered boulders provides a safe roost for a good number of terns, gulls and cormorants. The bird for which Kommetjie is best known is the distinctively stocky, subtly coloured **Antarctic Tern**, which can reliably be found here in small numbers from April to October. By early spring, shortly before undertaking their return flight across the southern oceans, the birds have often already attained their superb white, grey and deep red breeding dress.

The tern roost also includes **Swift** and **Sandwich Terns** all year round; **Common Terns** dominate during the summer. A handful of the threatened **Bank Cormorant** can usually be found on the rocks throughout the year, alongside much more common **Cape**, **Crowned** and **White-breasted Cormorants**. An assortment of waders is usually found pottering among the technicolour rock pools, including the resident **White-fronted Plover** and **African Black Oystercatcher**, as well as migrant **Ringed Plover**, **Turnstone**, **Common Sandpiper** and **Whimbrel**. Kommetjie is also a well-known sea-watching vantage point during the winter months (p.39).

Wildevoëlvlei, a largish lake nearby, was once home to several localized waterbird species including **White-backed** and **Maccoa Ducks**, and is easily accessible from the Imhoff's Gift housing development (take the signposted road north from the M65, a few kilometres east of Kommetjie). **Great Crested Grebe** and **Yellow-billed Egret** still occur here. In recent years, however, the lake has suffered heavily from blooms of toxic blue-green algae, resulting in a dramatic drop in bird numbers. Nonetheless, it is always worth stopping for a quick scan.

CAPE OF GOOD HOPE RESERVE

The rugged coastline and windswept moorlands of this spectacular reserve at the south-westernmost tip of Africa have attracted countless visitors since the Cape was first rounded by the Portuguese navigator Bartholomeu Dias in 1488, during the first successful quest for a spice route from Europe to the East. Now incorporated into the Cape Peninsula National Park, the Cape of Good Hope reserve is an essential destination if you're spending a few days in Cape Town, and is most famous for its striking landscapes, rich history, and botanical diversity – over 1 000 species of plant occur here, eleven of which are found nowhere else on earth. This is in sharp contrast to the low density of the birdlife – a scarcity which is, fortunately, more than compensated for by the quality of a few of the local specials. The reserve is probably the best place in the world to see the localized **Hottentot Buttonquail** (see box opposite), while other specials include **Cape Siskin** (p.33*), **Peregrine Falcon** and **Plain-backed Pipit**. The map on p.14 depicts all the sites we mention below.

The dramatic sea cliffs at Cape Point form the focus of most visits to the reserve. There is a shop and restaurant here, and visitors can take a short funicular ride up to South Africa's oldest lighthouse (built in 1860),

The windswept tip of the Cape Peninsula juts spectacularly into the Atlantic.

which offers sweeping views of the ocean, the cliffs, and the craggy length of the Cape Peninsula. Hardened birders will be consoled that, while fulfilling this essential tourist activity, they are also optimally positioned to see the endemic **Cape Siskin**. This rather inconspicuous canary can be seen anywhere in the scrub around Cape Point, the best areas being along the paths to the sea-cliff view sites (starting below the toilets on the Cape of Good Hope side of the parking lot) and along the precarious path to the lower lighthouse. Common birds here are **African Black Swift**, **Cape Robin**, **Grey-backed Cisticola** (curiously localized on the Peninsula and not even recorded at Kirstenbosch), **Karoo Prinia**, **Southern Boubou**, **Red-winged Starling** and **Cape Bunting**. Keep a look out for **Peregrine Falcon** (see box, p. 18) in this vicinity, although it can turn up anywhere in the reserve. The Cape Point and Cape of Good Hope promontories must rank among the best seawatching sites in the world (p. 39).

Watch out for the cheeky Chacma Baboons (*Papio ursinus*) here and elsewhere in the reserve, where the bolder individuals will enthusiastically jump onto or even into cars and steal food and other parcels from unsuspecting visitors. Habituated baboons often become dangerous and need to be destroyed, so please don't feed them.

The open plains of the reserve, as mentioned, are probably the best place in the world to see the localized **Hottentot Buttonquail** (see box), although finding this bird usually poses a considerable challenge. Take the second tar road to the right (opposite Kanonkop on the map provided at the entrance) when driving south from the entrance gate, and continue for about 2 km until you reach the only road to the left. Park at the gate just past this junction, and walk north to the broad, shallow valley dominating the landscape to the west of Sirkelsvlei. Concentrate your search on the plains dominated by low (less than 40 cm high) tussocky brown reed-like restio plants

HOTTENTOT BUTTONQUAIL

Hottentot Buttonquail (*Turnix (hottentota) hottentota*) is a poorly known yet distinctive form of buttonquail, often lumped as a subspecies of **Black-rumped Buttonquail**. It is restricted to the Fynbos Biome and occurs only from Cape Town to Port Elizabeth, unlike the more northerly subspecies of Black-rumped Buttonquail (*Turnix (hottentota) nana*) which is found in grasslands across Africa. Previously thought to be very rare, Hottentot Buttonquail has now been found to occur quite widely in mountain fynbos. The Cape of Good Hope reserve is the best place to search for this recalcitrant bird, and surveys have estimated that the reserve holds approximately 300–500 individuals. Despite being fairly common, it is difficult to observe, since most sightings are of birds flushing just in front of one's feet! Note that it lacks the obvious black rump of the grassland Black-rumped Buttonquail, and the males (with streaky, straw-coloured upperparts) are best distinguished from **Common Quail** by their much smaller size and more fluttering flight, without the short glides often made by the latter. The more richly coloured females are further distinguished by their buffy-orange upperwing coverts. It is also occasionally seen at Sir Lowry's Pass and at Potberg in the De Hoop Nature Reserve. Below is the only known photograph of Hottentot Buttonquail, here on its nest near Cape Town.

Hottentot Buttonquail occurs on the plains of the Cape of Good Hope reserve.

and low bushes. It will soon become distressingly evident that there is a vast amount of habitat here, and the dedicated birding group may have to spend a few hours on these windswept plains, with no guarantee of success! Birds are curiously scarce here, although **Cloud Cisticola** (p. 57*), **Grey-backed Cisticola**, **Grassbird**, **Clapper Lark**, **Grey-wing Francolin** and the **Orange-throated Longclaw** may be seen.

The scenic Atlantic beaches and rocky shores near Olifantsbos lie off the main tourist itinerary and are ideal for a stroll. A trail leads south to the rock-perched hull of the *Thomas T. Tucker*. The small beach just north of the parking area is an excellent spot for **Plainbacked Pipit**, and several are usually seen foraging just above the high-water mark – rather atypical habitat for a pipit. **African Black Oystercatcher** (p. 32*), **White-fronted Plover** and **Little Egret** also forage on these shores, and they are joined in summer by **Whimbrel**, **Turnstone** and **Sanderling**, **Swift**, **Sandwich** and **Common Terns**. **Cape**, **White-breasted** and **Crowned Cormorants** roost on the rocks.

The coastal thicket adjacent to the parking area supports **Fiscal Flycatcher**, **Cape Bulbul**, **Southern Boubou** and **Speckled Mousebird**. **Ostrich**, and Bontebok *(Damaliscus dorcas dorcas)* – a once critically endangered antelope endemic to the Cape – graze in the open near the parking area. Flowering patches here and elsewhere in the reserve attract large numbers of nectarivorous species, such as **Cape Sugarbird** (p. 33*), **Orange-breasted Sunbird** (p. 33*), **Malachite Sunbird** and **Lesser Double-collared Sunbird**. Rocky places in the reserve are worth searching for the likes of **Ground Woodpecker**, **Jackal Buzzard**, **Cape Rock Thrush**, **Cape Siskin**, **Familiar Chat** and the much rarer **Sentinel Rock Thrush**.

BOULDERS BEACH, SIMON'S TOWN

This area, site of the larger of the two mainland colonies of the endearing and globally threatened **African Penguin** (p. 32*), is comprised of secluded sandy beaches that nestle among imposing granite boulders bordered by dense coastal thicket. Over 900 pairs of penguins now breed here (see box), peering suspiciously from their shallow, sheltered burrows at their now considerable following of tourists.

Additionally, **Cape** and **Crowned Cormorants** roost on the offshore boulders, while **White-backed Mousebird**, **Southern Boubou** and **Bully Canary** inhabit the thickets through which the footpath to the colony passes.

Take plenty of camera film, and please don't forget to check under your car for lurking penguins before driving off!

Boulders Beach is on the southern edge of Simon's Town, and can be reached from the bottom of Bellevue Road (well signposted from Main Road: turn down along the northern edge of the golf course). A scenic harbour overlooking False Bay, Simon's Town has long served as a naval base (both British and South African) and has a rich maritime history. The harbour is a popular departure point for pelagic seabird trips (p. 38). There is a diversity of restaurants close to the seafront, conveniently situated for those wishing to elegantly round off a dedicated day on the southern Peninsula.

BOULDERS PENGUIN COLONY

The penguins seem so much at home at Boulders Beach that it is difficult to believe that they are relatively recent arrivals. A pioneering pair first nested here in 1985, launching a colonization process that has seen the colony expand rapidly. It is bolstered each year mainly by immigrant birds who desert their natal colonies for this secluded haven, free from natural predators (owing to human disturbance), and with good feeding grounds nearby (there is no large-scale commercial fishing in the adjacent bay). Breeding occurs throughout the year, with a distinct peak in the winter months. The eggs, usually two, are laid in protected environments – in burrows or under bushes – although ruthless competition for space forces many birds at Boulders Beach to nest in exposed sites.

While most of the area's human residents view the penguins as a natural asset, others are disturbed by the noise and smell of the colony, and by the increase in tourist traffic.

Boulders Beach, where African Penguins nest within metres of suburban villas.

STRANDFONTEIN SEWAGE WORKS

Although the uninitiated will often turn up their noses at the idea of voluntarily visiting a sewage farm, such places are often exceptionally rich in birdlife. This is especially true of the extensive Strandfontein sewage works, arguably the best waterbird locality close to Cape Town, whose existence is under threat from a new motorway. The abundant and diverse birdlife makes it an ideal destination for the beginner and serious twitcher alike, and it is possible to see more than 80 species on a summer morning. A major advantage is the opportunity to bird from the comfort and security of your car, which can be used as a moving hide. The vast network of reed-fringed pans which radiate out from the sewage plant buildings is connected by good gravel roads, but beware of occasionally treacherous sandy patches, especially along the southern coastal road.

To enter the Strandfontein sewage works from the Cape Town side, take the M5 freeway southwards from Cape Town and turn left into Ottery Road at the Ottery turn-off; continue for 4.5 km until the junction with Strandfontein Road (M17); turn right here, and continue (southwards) along Strandfontein Road for 4 km; turn right again at the 'Zeekoeivlei' sign (① on site map, opposite) within a stand of gum trees just after a petrol station and opposite Fifteenth Avenue. To enter the works from the False Bay side, turn north onto Strandfontein Road from Baden Powell Drive, 6.8 km east of the Muizenberg traffic circle, and you'll reach the Zeekoeivlei turn-off after 4.1 km.

Baden Powell Drive (R310) follows the False Bay coast westwards to Muizenberg and Simon's Town, and eastwards to the N2 highway near Somerset West. Strandfontein can thus conveniently be visited after Sir Lowry's Pass (p.60).

African Marsh Harrier

The poorly marked entrance to the works is adjacent to a derelict building at the south end of Zeekoeivlei (②), where **African Fish Eagles** are often seen roosting in the trees to the west. Bird numbers and water levels at Strandfontein vary widely depending on the

---

RECENT ARRIVALS ON THE PENINSULA

There is a host of species that, despite seeming so natural a part of the Cape avifauna, are recent colonizers of this far southwestern extremity of the continent. Why then the Cape's sudden popularity? Ironically, the answer lies in some of man's most ecologically destructive activities. Large bodies of fresh water were scarce in the southwestern Cape until farmers began to build dams, which have expanded both the range and numbers of many waterbird species. Among these are such familiar birds as **Sacred Ibis**, **African Spoonbill** and **Blacksmith Plover**, all of which were virtually unknown in the Cape until perhaps fifty years ago.

Then there is the introduction of alien trees, which has led to the demise of many splendid tracts of fynbos but appears also to have permitted the steady westward encroachment of a number of species more characteristic of the moister, more wooded east. In the past two decades several forest raptors, such as **Forest Buzzard** and **Black Sparrowhawk**, have found new breeding habitat here. Thus too have the **Acacia Pied Barbet** (with its brood parasite, the **Lesser Honeyguide**, hot on its tail) and **Red-eyed Dove** been lured to the Atlantic shoreline. The most conspicuous recent arrival has been that of the raucous **Hadeda Ibis**, almost unknown on the Peninsula just ten years ago but now a familiar sight and sound in Cape Town's suburbs.

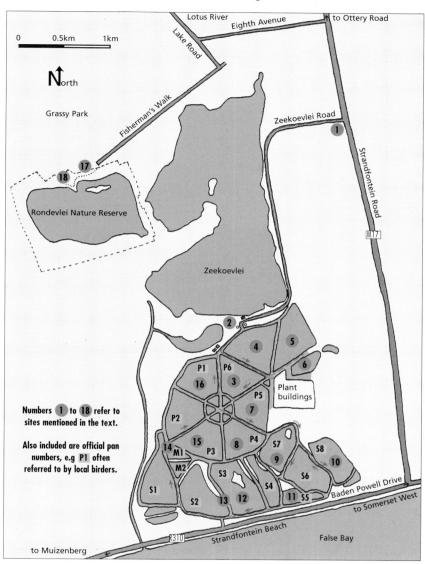

Numbers **1** to **18** refer to sites mentioned in the text.

Also included are official pan numbers, e.g **P1** often referred to by local birders.

year and season, and the route suggested below is intended as a general guide to the most productive areas.

Continue along the tar road towards the plant buildings, and check the deep pans on both sides of the road (③ and ④) for **Black-necked Grebe**, **Maccoa Duck**, **Southern Pochard**, and **Cape Teal**. Here too you will see the first of various other waterfowl species that are common throughout the sewage works, such as **Cape Shoveller**, **Yellow-billed Duck** and **Red-billed Teal**,

The bird-rich pans at Strandfontein sewage works always hold Greater Flamingo.

while **Purple Gallinule** stalk along the reed-lined edges. **Levaillant's Cisticola** is very common in long grass fringing the pans, and agitated birds draw attention to themselves with their characteristically frenetic calls. **White-throated** and **European Swallows** (summer) and **Brown-throated Martin** dart low overhead.

Where the road meets the sewage plant itself, continue to the left of the buildings, and scan pan ⑤ for a good variety of waterfowl. The adjacent small, muddy pan at ⑥ often host somewhat scarcer species such as **Southern Pochard** and **Wood Sandpiper**. The road between the two pans is regularly used in summer as a roost by large numbers of **White-winged Terns**, which can be seen flying over pans throughout the area.

At this point, retrace your route and continue to the pan at ⑦. This pan, and the small, reed-enclosed pond at its northern end, are usually also productive. At the 'hub' of the wheel of large pans, turn left. Pan ⑧, on your right, invariably holds good

numbers of birds, notably **Black-necked Grebe, White Pelican, Greater Flamingo** and **Maccoa Duck**.

The western and northern corners of pan ⑨ are always worth investigating. The former often has an exposed beach frequented by waders (including **Avocet**); the latter is good for scarcer ducks such as **Cape Teal** and **South African Shelduck**, and occasionally **Hottentot Teal**. Continue around the northern apex of pan ⑨ and head south past pan ⑩. The reeds in this vicinity are particularly good for **African Sedge Warbler, Cape Reed Warbler** and, in summer, **African Marsh Warbler**. Very much more evident in the alien thicket are **Cape Francolin** and **Cape Bulbul**. Pan ⑩ itself usually offers great birding, providing a good selection of waterfowl and wading birds in its northern reaches.

Options are now limited by sandy roads, so we suggest that you retrace your route and turn left along the southern border of pan ⑨. This is an especially good area for

African Marsh Harrier, which is virtually guaranteed to be seen flying low over the alien thicket and adjacent reedbeds. Head south again, and cast a glance over pan ⑪ for **African Black Oystercatcher**. Turn right where the road meets the coastal dunes, where **Swift** and **Sandwich Terns** and **Little Stint** (summer) often roost. Spare a moment to look up from your telescope and enjoy the splendid view over False Bay and its embracing mountains!

Good numbers of waterbirds can reliably be found on pan ⑫. **Cape** and **White-breasted Cormorants**, **White Pelicans** and miscellaneous waterfowl roost on the large, sandy island and on the pan edge (⑬ on map), while rafts of assorted ducks bob on the usually choppy water. A pair of **South African Shelduck** often frequents this pan, as do flocks of **Greater Flamingo**.

Having absorbed all pan ⑫ has to offer, continue past a series of relatively unexceptional pans before re-entering the central wheel at ⑮. The small pan at ⑭ is often productive, as is ⑮. Before leaving, you might find it worthwhile to check pan ⑯ for **Great Crested Grebe.**

RONDEVLEI NATURE RESERVE

Rondevlei and its surrounding reedbeds are protected in this cosy, well-maintained reserve, where Hippopotamus *(Hippopotamus amphibius)* has been reintroduced. The wide selection of common bird species, the excellent series of hides along a short, pleasant footpath, and the small museum make it especially suited to the beginner. Interesting species include **Darter**, **Little Bittern** (scan the reedbeds from hides at ⑱ on site map, p.27), **Purple Heron**, **Purple Gallinule**, **Painted Snipe** (rare), **Ethiopian Snipe** and a selection of waders (when water levels are low), and **Acacia Pied Barbet** (in the surrounding bush). Rondevlei's entrance (⑰) is situated at the western end of Fisherman's Walk, and is conveniently reached from the nearby Strandfontein sewage works.

Purple Gallinule – easily seen from the hides at the Rondevlei wetland.

Rondevlei has an excellent selection of bird hides and observation towers.

EXPLORING THE CITY

First port of call is the Cape Town Tourism offices in Burg Street, (see Useful Contacts, p.136), where advice, assistance and an assortment of helpful maps are on offer.

Cape Town's leading tourism development is the Victoria & Alfred Waterfront. Safe, sophisticated, and boasting an unrivalled choice of retail and speciality outlets, restaurants, cinemas and hotels, the Waterfront is just the place for the birding-weary spouse and family. Boat trips to Robben Island (below) depart from here. Near the entrance is the Two Oceans Aquarium, a world class facility with numerous well-constructed displays of South Africa's marine ecosystems. Also enquire about the World of Birds bird-park (wide selection of local and foreign birds) in Hout Bay and the South African Natural History Museum in Queen Victoria Street.

The best spots close to the city to enjoy the sunset after a hard day's birding are along the Atlantic coast, especially the trendy seaside suburbs between Sea Point and Camps Bay. Here you'll find a fine selection of beachfront walks, restaurants, and rocky perches for sundowners.

ROBBEN ISLAND AND SEA POINT

As the place where Nelson Mandela spent many of his prison years, Robben Island is burdened with a notoriety disproportionate to its small size and unprepossessing appearance. However, for birders, it is renowned not only for its sinister political history: the island supports significant seabird breeding colonies, including a substantial population of the endemic **African Penguin** (p.32*), and is of additional local interest in that it plays host to two introduced species found nowhere else in South Africa. The prison, including Nelson Mandela's cell, are visited as part of the 3-hour organized tour, including ferry transport, that is currently the only way to visit the island. Ferries depart from the Waterfront hourly from 09h00 to 12h00, and at 14h00.

Robben Island, once a notorious prison, hosts breeding seabirds and introduced species.

In 1964, customs officials in Cape Town used the island as a conveniently isolated depot for half a dozen captive-bred **Chukar Partridges**, which are native to Europe and Asia. The birds have since flourished, and small coveys are usually seen on the prescribed bus tour as they scurry through the alien thicket and low scrub that covers much of the terrain.

Found lurking in the denser thicket is the other introduced species that gives the island its birding reputation – **Common Peafowl**. Though many small feral populations of this familiar species exist in South Africa, only the Robben Island birds have been officially recognized as genuinely wild-breeding, thus legitimately worthy of listing.

Of rather higher significance in a global context are the substantial breeding colonies of **African Penguin**, **Bank Cormorant** (breeding on the harbour breakwater; see box, p. 21), **Crowned Cormorant**, **African Black Oystercatcher** (p. 32*), **Hartlaub's Gull** and **Swift Tern**. These are all easily seen along the island's coastline. A boardwalk offers access to the penguin colony. There can also be interesting seabirding en route to and from the island, and **Sabine's Gulls** may be seen lifting off the waves on tri-coloured wings as the ferry ploughs its way across Table Bay in summer. **Arctic** and **Pomarine Skuas** can also been seen in summer, while in winter **Subantarctic Skuas** mercilessly harry the other seabirds for their hard-won meals.

For those with limited time, or less of a stomach for the choppy ride across the bay, there is plenty to see along the city's western seaboard. The alternately rocky and sandy shoreline from the Waterfront westward to the suburbs of Mouille Point and Sea Point supports small numbers of roosting **African Black Oystercatchers**, **Cape** and **Crowned Cormorants**, as well as **Swift** and **Common Terns**. South of Sea Point, the coastal road passes through the haunts of Cape Town's wealthiest and trendiest, the suburbs of Camps Bay, Clifton and Bantry Bay.

INTRODUCED BIRDS OF CAPE TOWN

The Western Cape has the dubious distinction of hosting the country's greatest diversity of alien bird species. The 'usual' ones that have colonized much of the world (such as **House Sparrow**, **Feral Pigeon** and **European Starling**) are of course present, and there is an additional assortment of others that have become heavily twitched by list-conscious South African birders. Of these, **Mute Swan** has become locally extinct (although a wandering individual is occasionally seen at Dick Dent Bird Sanctuary on the R44 in Strand), **House Crow** (now distressingly well-established on the Cape Flats: look out for it on the N2 near the airport turn-off), **Peafowl**, **Chukar Partridge** (opposite) and **Chaffinch**. The latter is the only surviving relic of Cecil John Rhodes's 1898 bout of introductions, part of a broader attempt to transform the Cape Peninsula into a gentle English landscape. Among his other, less successful importations were **Rook**, **Song Thrush** and **Blackbird**. The **Chaffinch**, however, is peculiar in that it has neither gone extinct, nor become invasive, but remains peacefully ensconced in densely planted areas on the eastern slopes of Table Mountain. It is fairly common, although rather elusive and best lured down from the tree-tops by playback. Good areas to look (and listen) for it are Tokai (p.20) and the Greenbelts (p.19). **Mallard** is still fortunately fairly scarce, and regular reports of hybridization with **Yellow-billed Duck** are disturbing.

At Bakoven, good numbers of the globally threatened **Bank Cormorant** (see box, p. 21) breed on the elephant-like boulders that lie just offshore, and are best observed by telescope.

SELECT SPECIALS

African Penguin

Formerly known as the Jackass Penguin because of its loud, peculiarly braying call, this charismatic bird has acted as a mascot species for marine conservation in South Africa. Although two mainland colonies have recently formed near Cape Town (see pp.24 and 62), African Penguins typically breed on offshore islands from Namibia to the Eastern Cape

Province. The past century has seen a shattering 90 per cent loss of the population, which now stands at a steadily declining 160 000. Old enemies have been replaced by new ones: egg collecting and guano harvesting once took a heavy toll on the breeding colonies; today their main food source, pilchards, is threatened by over-fishing. Oil spills also cause devastating losses, although the South African National Foundation for the Conservation of Coastal Birds (SANCCOB) has over the past two decades successfully rescued and rehabilitated many tens of thousands of oiled birds (see p.43).

African Black Oystercatcher

Fewer than 4 800 individuals of this striking endemic still grace rocky and sandy shores from Namibia to the Eastern Cape Province. Although it is one of the world's rarest oystercatcher species, it is conspicuous on the Cape Peninsula and

can easily be seen at Sea Point (p.31), Kommetjie (p.21) or in the Cape of Good Hope reserve (p.24). In fact, more than half the world population occurs within 300 km of Cape Town. The birds' demise lies in their habit of nesting on exposed beaches: the superbly camouflaged eggs and young chicks are often trodden on or driven over. Professor Phil Hockey of the Percy FitzPatrick Institute of African Ornithology leads a long-term study of the ecology and conservation of this species.

Knysna Warbler

This localized endemic, which occurs along the south coast from Cape Town to the Eastern Cape, is best searched for in the riverine undergrowth of the forests and thickets of the eastern slopes of Table Mountain (see Constantia Greenbelts, p.19, and Kirstenbosch, p.17). There are also good sites at the Grootvadersbosch Nature Reserve (p.69) and on the Garden Route (p.117). It is an extreme skulker, though its penetrating, descending song makes up for its admittedly drab plumage and secretive habits. Its low, rattling contact calls are

rather less conspicuous, but are often useful in locating it. Stealthy and mouse-like, it spends much of its time feeding on the ground, flicking its wings and tail to disturb insects from the leaf litter. Despite occurring in the city suburbs, its nest was only discovered as recently as 1960, in the Kirstenbosch National Botanical Garden.

Cape Sugarbird

There are just two species of sugarbird, together constituting southern Africa's only endemic bird family. Their puzzling evolutionary relationships continue to perplex scientists, with conflicting evidence variously suggesting relationships with the starlings, the sunbirds, or even the Australian honey-eaters. The Cape Sugarbird is easily seen, conspicuous in distinctive silhouette, as it perches on the flowerheads of the protea bushes that are a major element of fynbos vegetation. Local sugarbird populations fluctuate markedly over the year, with the birds ranging up to 150 km in search of flowering protea bushes. Nonetheless, sugarbird-bedecked flowering proteas are found year-round in the Kirstenbosch National Botanical Garden (p. 17); other good sites are the Cape of Good Hope reserve (p. 24), Sir Lowry's Pass (p. 61) and Swartberg Pass (p. 123).

Orange-breasted Sunbird

Another fynbos endemic, the Orange-breasted Sunbird is primarily a specialist feeder on the spectacularly diverse, often tubular, erica (heath) flowers that are a major component of fynbos. It is often the most abundant species in this habitat, and is easily located by its characteristically metallic call, one of the distinctively atmospheric sounds of the rugged, wind-battered Cape mountains. Sunbirds at Kirstenbosch (p. 17) are often very approachable, but this species is also easily found at the Cape of Good Hope reserve (p. 24), Sir Lowry's

Pass (p. 61), Harold Porter Botanical Garden (p. 63), Bain's Kloof Pass (p. 81), Swartberg Pass (p. 123), or indeed any expanse of mountain fynbos.

Cape Siskin

One of the trickier fynbos endemics to find during a short visit, the Cape Siskin is nonetheless often a common bird in its preferred habitat, which is open, rocky fynbos, usually wind-blown and not conducive to good views of a small, mobile seed-eater. Cape Siskins are usually detected as they give their distinctive tri-syllablic flight call before disappearing over a ridge in small, nervous flocks. The males' white-tipped wings and tail are a good field character. For details of the best sites, refer to the text on the Cape of Good Hope reserve (p. 81), Bain's Kloof Pass (p. 81), Grootvadersbosch Nature Reserve (p. 70), Harold Porter Botanical Garden (p. 63) and Swartberg Pass (p. 123).

Seabirding

'And a good south wind sprung up behind; The Albatross did follow...'
SAMUEL TAYLOR COLERIDGE, *THE RIME OF THE ANCIENT MARINER*

Approaching a fishing trawler off the continental shelf near Cape Town is a birding experience that will remain engraved in your memory forever. Having thousands of albatrosses only an arm's length away is a highlight of most birding trips to Cape Town. For first-timers, it is an awesome spectacle, and for more experienced hands it's a great chance to search for adrenaline-pumping rarities. Day trippers in winter regularly see over 10 000 seabirds of up to 30 species, making it arguably the world's most memorable yet easily accessible sea-birding experience.

Pelagic birding off Cape Town.

The Cape's amazing seabird abundance and diversity is the product of the Benguela current that originates in the icy waters of Antarctica. Surging up the west coast of southern Africa, the nutrient-rich waters cause upwelling along the continental shelf, nurturing a profusion of ocean life that supports both a lively fishing industry and vast numbers of seabirds. Pelagic species – those which breed on land but otherwise remain at sea – congregate around the trawlers, making them easy to locate and approach. The high point of a pelagic birding trip is sure to be that of wallowing behind a trawler with up to 5 000 birds squabbling for scraps in its wake.

Most trips leave at dawn from Simon's Town harbour, on the southern Peninsula, and head south past the towering cliffs of Cape Point, a fine sight as the sun comes up over False Bay. As you round the tip of the Cape Peninsula, you will quickly feel the rolling Atlantic swells, heightening the anticipation of potential birds to come. The diversity of seabirds is highly seasonal, so consult the monthly table (opposite), compiled from over 300 pelagic birding trip lists from the past 10 years. This will help you to decide when best to go in order to maximize the chances of seeing your most-wanted species. Seabird numbers do fluctuate from year to year, and birding in the vicinity of a trawler will make a huge difference to your list.

WINTER TRIPS

Winter (May to September) is the most spectacular time at sea. Huge numbers of albatrosses and other pelagic seabirds migrate northwards from their breeding

SEABIRDS OFF THE CAPE (opposite)
A seasonal table for all regularly occurring species based on over 300 pelagic seabird trips over the last 10 years. Colour codes refer to the percentage chance of encountering each species. The authors gratefully acknowledge Barrie Rose's assistance in preparing this table.

SEE BOX OPPOSITE, BELOW   Key: 5–25% ░   25–50% ▒   50–75% ▓   75–100% █

	JAN	FEB	MAR	APR	MAY	JUN	JUL	AUG	SEP	OCT	NOV	DEC
Northern Royal Albatross					░	▓	▓	▓	░	░		
Wandering Albatross					▒	▓	▓	▓	▓	░		
Shy Albatross	█	█	█	█	█	█	█	█	█	█	█	█
Black-browed Albatross	█	█	█	█	█	█	█	█	█	█	█	█
Grey-headed Albatross					░	▒	▒	▒	░			
Yellow-nosed Albatross (total)	█	█	█	█	█	█	█	█	█	█	█	█
subsp. *chlororhynchus*	█	█	█	█	█	█	█	█	█	█	█	█
subsp. *bassi*	░	█	█	█	█	█	█	█	█	█	█	█
Southern Giant Petrel	░	░	░	░	▒	█	█	█	█	▒	░	░
Northern Giant Petrel	░	▒	░	░	▒	█	█	█	█	▒	░	░
Antarctic Fulmar						░	░	░	░			
Pintado Petrel					▒	█	█	█	█	▓	░	
Great-winged Petrel	█	█	█	█	█	░	░	░	░	█	█	█
Soft-plumaged Petrel					▒	▒	▒	▒	▒	░		
Antarctic Prion				░	▒	█	▒	░				
White-chinned Petrel	█	█	█	█	█	█	█	█	█	█	█	█
Spectacled Petrel	░	░	░	░	░	░	░	░	░	░	░	░
Cory's Shearwater	▓	▓	▓	▓	▒	░					▓	▓
Great Shearwater	▓	▓	▓	▓	▒	░		░			▓	▓
Flesh-footed Shearwater	░	░	░	░	░					░	░	░
Sooty Shearwater	█	█	█	█	█	█	█	█	█	█	█	█
Manx Shearwater	▒	▒	▒	▒						▒	▒	▒
European Storm Petrel	█	█	█	█					░	▒	█	█
Leach's Storm Petrel	░	░	░	▒							▒	░
Wilson's Storm Petrel	▓	▓	▓	▓	▒	▒	▒	▒	▒	▒	▓	▓
Black-bellied Storm Petrel					░				░	▒		
Cape Gannet	█	█	█	█	█	█	█	█	█	█	█	█
Arctic Skua	▓	▓	▓	▓	▒	░			░	▓	▓	▓
Long-tailed Skua	░	░	░	▒	░						░	░
Pomarine Skua	▒	▒	▒	▒	░					░	▒	▒
Subantarctic Skua	▓	▓	▓	▓	▓	▓	▓	▓	▓	▓	▓	▓
Sabine's Gull	▓	▓	▓	▒					░	▓	▓	▓
Arctic Tern	▓	▓	▓	▓	░		░	▒	▒	▓	▓	▓

Above: Rounding Cape Point at dawn.
Opposite, below: Many thousands of pelagic seabirds are attracted to trawlers.

sites as far south as Antarctica, moving into Cape waters to escape the harsh polar winter. **Shy** and **Black-browed Albatrosses** are abundant, and both subspecies of **Yellow-nosed Albatross** are commonly seen in small numbers. The waters off the Cape are the most accessible place globally to see the grey-headed *chlororhynchus* subspecies of the latter, which is often regarded as a full species (see box, opposite). The great prize of a winter trip must however be the romantically celebrated **Wandering Albatross** (p.40★), although the spellbinding star of Samuel Coleridge's *The Rime of the Ancient Mariner* has become very scarce in recent years (see box, p.39). There is also always a chance of seeing the rare **(Northern) Royal** and **Grey-headed Albatrosses**.

The ever-present **White-chinned Petrel**, **Sooty Shearwater** and **Cape Gannet** are joined by huge numbers of flashy **Pintado Petrel**, **Broad-billed Prion** (*desolata* subspecies is most common) and **Wilson's Storm Petrel**. Both the **Northern** and **Southern Giant Petrels** (p.40★) are invariably present in small numbers, usually one or two per trawler, and **Antarctic Fulmar** and **Spectacled Petrel** make an occasional appearance. Watch out for the odd fast-flying **Soft-plumaged Petrel** whipping by, especially away from the trawlers. Small flocks of terns fly by, and **Subantarctic Skua** is usually present at each boat and is often seen even before leaving False Bay. **Antarctic Tern** are sometimes seen close inshore.

SUMMER TRIPS

From October to April, North Atlantic seabirds migrate south to claim their share of the Benguela's bounty. Although seabird numbers are generally smaller during this period than in winter, summer trips are spectacular nonetheless and do provide an opportunity to see several additional species.

The three most common albatrosses, the two giant petrels, and **White-chinned Petrel** and **Sooty Shearwater** are, as always, present. They are joined in summer by **Cory's** (mainly *diomedea* subspecies, also known as **Scopoli's Shearwater**) and **Great Shearwaters**, along with smaller numbers of **Soft-plumaged Petrel** (early summer) and the occasional **Spectacled Petrel**, and **Manx** and **Flesh-footed Shearwaters**. The majority of **Great-winged Petrels** make their appearance only at this time, having spent the harsh Antarctic winter at their breeding grounds further south. Good numbers of **European Storm Petrel** join the ever-present **Wilson's Storm Petrel**, while the rare **Leach's Storm Petrel** (p.40★) is mainly seen beyond the continental shelf. **Black-bellied Storm Petrel** is present only in small numbers, as a passage migrant, in late September/October and again in April.

The common **Arctic**, uncommon **Pomarine** and **Subantarctic** and very scarce **Long-tailed Skuas** patrol the skies closer inshore. **Arctic Tern** is a passage migrant, although it is seen in smaller numbers throughout the summer, along with **Sabine's Gull**. Lucky observers may see small flocks of **Grey Phalarope**.

RARITIES AND THE 1984 SEASON

For local birders, it is the lure of local rarities that makes the pelagic trips so popular. Almost anything can turn up, including the following, characteristically in winter, species recorded in the Western Cape: **(Southern) Royal Albatross** (3 records at sea), **Buller's Albatross** (1 record at sea), **Dark-mantled** (2 confirmed records at sea, 2 on land, 10 washed up dead on beaches) and **Light-mantled Sooty Albatrosses** (2 confirmed records at sea, 3 beached), **Antarctic Petrel** (2 beached), **White-headed Petrel** (2 at sea, 1 beached), **Atlantic Petrel** (very scarce, no figures available), **Kerguelen Petrel** (very scarce, except in 1984), **Blue Petrel** (very scarce, except in 1984), **Slender-billed Prion** (very scarce, except in 1984), **Fairy Prion** (1 beached), **Grey Petrel** (very scarce), **Little Shearwater** (scarce), **Black-legged Kitti-wake** (2 at sea, 1 on land) and **South Polar Skua** (scarce). Rarities seen in summer include **White-bellied Storm Petrel** (very scarce) and **Laysan Albatross** (1 at sea).

In July 1984, a remarkable seabird irruption occurred from South Africa to faraway Australia and New Zealand. This was possibly linked to the El Niño weather conditions prevailing during the previous season, and was associated with many beached seabird corpses. There were sightings of birds ordinarily very rare at sea, including large numbers of **Kerguelen Petrel**, **Blue Petrel** and **Slender-billed Prion**. The most bizarre record was surely that of the dazed **Dark-mantled Sooty Albatross** found atop an apartment block in suburban Cape Town!

CHANGING ALBATROSS TAXONOMY

Recent and controversial research on albatross taxonomy, based on genetic analysis and embracing the phylogenetic species concept (p.13), has suggested that there are ten unrecognized species of albatross in the world. These tentative 'new' species are currently classified as subspecies, but should they be recognized as full species, the global albatross species total would rise to from 14 to 24. In the seabird seasonality table, the proposed 'new' species that can be distinguished at sea are treated separately should they be split in the future, such as the *chlororhynchus* subspecies of **Yellow-nosed Albatross**, shown here.

37

The new lighthouse near the tip of Cape Point: one of the world's best seawatching spots.

ORGANIZING A PELAGIC TRIP

Reasonably priced day trips, led by experienced local leaders, depart from Simon's Town harbour about once a month (more often in winter). It is also possible to charter more speedy and luxurious private game-fishing vessels departing from Hout Bay. Please contact the authors (see Useful Contacts, p.136) for up-to-date pelagic birding trip information.

Pelagic birding vessels use radar to detect and approach trawlers on the continental shelf, 30–40 km offshore. Ensure that you arrive early and get good directions to exactly where to park in the harbour. Conditions may turn fairly rough, especially during winter, so be sure to bring anti-nausea tablets if you are prone to seasickness. If you do feel sick, try to stay outside in the fresh breeze, and keep your eyes on the horizon. If you can, stay near the middle of the boat where there is a little less motion, and try to avoid the diesel fumes, which only exacerbate one's malaise. Conditions on board can vary substantially from quite hot to exceptionally cold, particularly on rough days when sheets of spray may soak one at intervals. Essential items include a waterproof anorak, a jersey, some spare clothing, and a woolly cap or hat which won't get blown off. Sunglasses and sun-screen are essential for the glare (don't underestimate the sun offshore – you will get quite burnt even when the air is icy). It is of course worth bringing a camera, but be sure to protect it from the salt water.

SEAWATCHING FROM THE PENINSULA

Those who don't trust their sea legs may consider taking their telescopes out on a windy day and gazing out to sea to search for pelagic seabirds that are blown inshore. Although the popularity of this pastime has declined recently due to the increased availability of pelagic birding trips, there are still some sites worth visiting on the Peninsula if you are a hardened seawatcher or a weakened seafarer.

In winter, seawatching is best on the western side of the Peninsula when a strong northwesterly is blowing. Try to find a position elevated enough to preclude your quarry dipping infuriatingly behind the wave troughs, and if possible sheltered from

THE LONGLINE KILLER

Research is demonstrating that the fairly recent advent of longline fishing techniques is causing a tragic number of deaths among southern hemisphere seabirds. Fishing lines up to 100 km long, studded with up to 20 000 baited hooks, are trailed behind fishing vessels. It is estimated that a staggering 100 million hooks each year are set in the southern oceans alone. As the line is lowered into the water, but before it sinks very deep, seabirds following the boat plunge down to grab the bait, get hooked and drown. Research estimates suggest that as many as 40 000 albatrosses are killed annually, a disturbing figure which is causing population declines in several species. These declines are potentially devastating, especially among the long-lived Wandering Albatrosses (p.40*), a species which only raises one chick every two years. Currently, the Global Seabird Programme of BirdLife International and other concerned parties are investigating ways of reducing this seabird mortality.

light rain squalls. The best spots are at the Cape of Good Hope (find a sheltered vantage point on the cliffs above the parking area; see map p.14), Cape Point (take the path from the old lighthouse to the new one; p.22) and Kommetjie (from the shore near the lighthouse; p.22). Even the casual seawatcher is bound to see a sprinkling of **Cape Cormorant**, **Cape Gannet**, **White-chinned Petrel** and **Sooty Shearwater** just offshore. If there is a strong wind, **Shy** and **Black-browed Albatrosses** may also be seen, with regular appearances made by **Sub-Antarctic Skua**, **Northern** and **Southern Giant Petrels**, **Yellow-nosed Albatross**, **Wilson's Storm Petrel** and **Broad-billed Prion**.

In spring, summer and autumn, the persistent southeasterly winds produce good seawatching, and the best vantage points are Glencairn (made famous by seawatching expert, Dr Mike Fraser) and Cape Point. Glencairn is a small suburb on the east coast of the Peninsula, between Fish Hoek and Simon's Town. Stand next to the railway station, or at the whale-watching site 1 km north of the railway station. The seawatching is best in spring and late summer (October and February–March) on the first or second day of the southeaster. Birds are blown into False Bay and are best viewed in the late afternoon as they move south, out of the bay. Most common are **Cape Gannet**, **Arctic Skua**, **Sooty Shearwater** and **White-chinned Petrel**. Less common but regular nonetheless are **Pomarine Skua** and **Cory's Shearwater**; scarcer still are **Soft-plumaged Petrel**, **Great Shearwater** and **Long-tailed Skua**.

In summer, scan offshore from the Mouille Point lighthouse (just west of the V&A Waterfront, see p.31) for distant flocks of **Sabine's Gull** (October–April), as well as **Cape Gannet**, **White-chinned Petrel**, **Arctic Skua** and **Swift Tern**.

SELECT SPECIALS

Wandering Albatross

Few sights epitomize the freedom of the open oceans as elegantly as a soaring Wandering Albatross. With the longest wingspan of any bird (in excess of 3.5 metres in some cases), these majestic ocean travellers are able to spend many months on the open sea, effortlessly exploiting updrafts from the waves to stay aloft.

They are perhaps most famous for their life-long pair bonding display, during which the two birds face each other with wings outstretched and bills pointing skywards. However, this long-lived star of many nature documentaries is under threat, and it is believed that up to 10 per cent of the world population may be lost to longlining each year (p. 39). As pairs can only raise, at most, one chick every two years, and because it will take 11 years before the offspring is ready to breed, urgent steps must be taken to avert the imminent extinction of this graceful seafarer.

Spectacled Petrel

The 'Ringeye', as it is more affectionately known, was only recently recognized as a full species, split from White-chinned Petrel. This taxonomic decision, based largely on the breeding calls, bestows upon it the dubious distinction of being one of the world's most threatened seabirds. Only about 10 000 individuals exist, breeding only on Inaccessible Island in the South Atlantic Ocean. Alarmingly, it is believed that as much as 5 per cent of the population is killed annually by longline fishing off Brazil (p. 39). The diagnostic white facial crescent separates it from White-chinned Petrel only at close range, and care must thus be taken not to confuse it with occasional White-chinned Petrels that show white patches on the head.

Leach's Storm Petrel

Because it is regularly seen only far offshore in Cape waters, Leach's Storm Petrel was assumed to be an exclusively non-breeding migrant from the northern hemisphere during our summer months. However, this species was discovered as recently as 1997 by Phil Whittington to be breeding on Dyer Island (near Hermanus; see map, p. 58), making it the African continent's only breeding pelagic seabird. Up to twenty pairs of birds breed on the island annually, and can be heard calling at night from their nesting burrows deep in the old stone walls that surround the island's few buildings. The lateness of this discovery can be attributed to the birds' strictly nocturnal activity.

Southern Giant Petrel

Because it is difficult to distinguish from its sister species, the Northern Giant Petrel, many giant petrels seen in Cape waters remain indeterminate. The most useful feature in separating the two species is the bill tip: in the Southern Giant Petrel it is a greenish colour, while in the Northern Giant Petrel it is horn-coloured. Unique to the Southern Giant Petrel is the rare white phase, in which the whole bird is an ivory colour. The giant petrels are the vultures of the sea, often scavenging on dead seals, especially on their breeding grounds. A century ago, giant petrels used to gather in great numbers to scavenge at the Cape's whaling stations.

West Coast

'The gulls' gab and rabble on the boat-bobbing sea ... scamper of sanderlings, curlew cry ... he got a little telescope to look at birds...'. DYLAN THOMAS, UNDER MILK WOOD

The southwestern Cape's western seaboard, stretching along the Atlantic shores from Cape Town northwards to the Olifants River, is best known for its superb beaches, bountiful sealife, internationally recognized coastal wetlands, and spring wildflower displays that are nothing short of spectacular. Birding is excellent: there is an abundance of migrant waders and other waterbirds, and rewarding 'strandveld' birding. Highlights range from the quiet elegance of a Black Harrier quartering low over the scrublands of the West Coast National Park, to the frenzied activity of the Cape Gannet colony at Lambert's Bay.

This topographically unassuming region is dominated by a coastal plain, covered in low, scrubby strandveld vegetation and studded in many areas with picturesque granite outcrops. Distinctly different in character from the Cape's southern seaboard, the West Coast is decidedly more arid and exposed, with the scrublands offering little protection

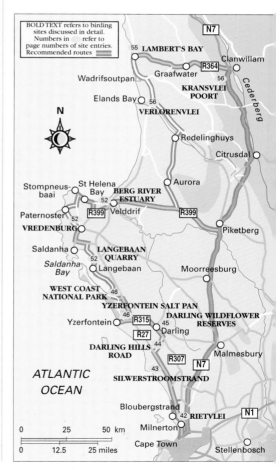

Cape Gannets at Lambert's Bay

TOP BIRDS
Black Harrier, Grey-wing Francolin, Southern Black Korhaan, Chestnut-banded Plover, Cape Long-billed Lark, Clapper Lark, Southern Grey Tit, Sickle-winged Chat, Cape Penduline Tit, Cloud Cisticola, Protea Canary, and a host of waders.

from the unrelenting sun and blustery onshore winds. Further inland, despite most of the fertile soils of the Swartland lying under intensive wheat cultivation, the birding is still remarkably productive. The rich waters of the Benguela Current not only make this region the heart of the country's fishing industry, but the associated sealife supports massive seabird breeding colonies on the scattered offshore islands. The coastline consists largely of endless lonely beaches, punctuated by salty, whitewashed fishing villages and an ever-increasing number of holiday retreats.

The West Coast is best birded in spring and early summer (from about August to October), when most of the resident birds are breeding and the wildflowers are at their peak. This contrasts strongly with late summer, when the region is particularly dry and many of the temporary waterbodies have evaporated, leaving arid depressions populated only by dust-laden whirlwinds. The southern areas of the West Coast, extending northwards to the West Coast National Park and even the Berg River estuary, can be comfortably explored in a day-trip from Cape Town. However, a two- to three-day loop would allow for more relaxed exploration of the region, including the Lambert's Bay area. The West Coast can also conveniently be visited en route to Bushmanland or Namaqualand. Birding is best in the mornings as it is usually persistently windy later in the day.

LEAVING CAPE TOWN: WATERBIRDS AND RIETVLEI

The main arterial road running up most of the western shoreline is the R27, which can be reached from the N1 national road just north of Cape Town. It begins in the coastal suburb of Milnerton, where it is initially known as Otto Du Plessis Drive. As you pass through Milnerton, scan the

lagoon that lies between the R27 and the conspicuous Woodbridge Island lighthouse on your left. A number of widespread waterbird species are often found here, most notably **Little Egret**, **Grey-headed Gull** (uncommon), **Caspian** and **Swift Terns**, and **Pied Kingfisher**.

Continuing a few kilometres further north, the road swings to the left. Here, a number of large waterbodies are visible on your right. Initially, the non-perennial pan of Rietvlei (dry and dusty in late summer) can be seen in the distance. During the winter and spring, the pan supports an excellent diversity and abundance of waterbirds, and these can be viewed from the bird hide on the opposite side of the pan (see below for directions). Birders visiting at the right time of year and with time to spare will find a visit to Rietvlei rewarding. However, most of its birds are more conveniently found elsewhere on this route. Continuing along the R27, the deep waters of Flamingo Vlei come into view on your right after a short while. This lake is used mainly for watersports and birdlife is less diverse, although **White Pelican** and **Darter** can often be seen here.

At the third set of traffic lights beyond the lighthouse (after 6.4 km), there is a series of pans surrounding the road: one lies to the left (Pan 1), another to the right (Pan 2), and one to the left beyond the traffic lights (Pan 3). These 'Dolphin Beach' pans can be birded from the roadside, and support a remarkable diversity of waterbird species, including **Dabchick**, **Yellow-billed Egret**, **Glossy Ibis**, **Cape Shoveller**, **Yellow-billed Duck**, **Red-knobbed Coot**, **Moorhen**, **Purple Gallinule**, **Ethiopian Snipe**, **Three-banded Plover**, **Black-winged Stilt**, and in summer, **Little Stint**, **Wood Sandpiper**, **Ruff** and **White-winged Tern**. The localized **White-backed Duck** is invariably present on Pan 3. The scarce and very local **Painted Snipe** is also occasionally found here, especially in the grassy edges of Pan 1. Scan the reedbeds for **Cape Reed Warbler**, and the vegetation along the edges of the

Above: Strandveld vegetation surrounds Langebaan Lagoon in the West Coast National Park.
Opposite: Lucky observers may see Painted Snipe near Rietvlei.

pans for the conspicuous **Levaillant's Cisticola** and **Common Waxbill**. **Brown-throated Martin** and **White-throated Swallow** hawk insects overhead, and **African Marsh Harrier** can often be seen over the reedbeds.

To enjoy panoramic frontal views of Table Mountain, turn left at the traffic lights towards Bloubergstrand. Check the rocks along the beach here for **Crowned Cormorant** and **African Black Oyster-catcher** (p.32*).

Returning to the R27, continue northwards. Should you wish to visit Rietvlei, turn right at the first set of traffic lights (after 0.7 km) beyond the roadside pans, take first right again (Pentz Drive) and continue for just over a kilometre until you reach the SANCCOB seabird rehabilitation centre on your right. This very worthwhile organization deserves a quick visit as there are always recovering seabirds on site (see p.32). From SANCCOB, turn right at the first four-way stop and inquire at the Aquatic Club for access to the bird hide. There is a 15 minute walk to the hide, and a small fee is payable.

ENTERING THE STRANDVELD: SILWERSTROOMSTRAND

Uniform thickets of tall vegetation line the road as you head north along the R27,

beyond the ever-expanding borders of Cape Town. These alien trees, introduced from Australia in the 19th century to stabilize the dunes, continue to spread and smother the indigenous plants. Resisting most attempts to root them from the landscape, they currently pose one of the greatest threats to the natural vegetation. As you proceed further north, the low, scrubby plant cover native to the area becomes more evident. Here referred to broadly as 'strandveld' (although it contains elements of lowland fynbos, see p.7), it occurs on very sandy soils and is characterized by low, dense, thicket vegetation (with many fruit-bearing shrubs) interspersed with stands of restios (brown, reed-like plants), and supports a rich bird community.

Interestingly, the prominent hill on the right, close to the R27, is the site of a battle that had a decisive influence on the course of history in South Africa. It was after this confrontation, nearly two centuries ago, that an invasion force of 63 British warships captured the Cape and brought 150 years of Dutch rule at the southern tip of Africa to an abrupt end. On the left, only 30 km from the centre of Cape Town, loom the controversial reactor domes of South Africa's only nuclear power station, Koeberg. **Black-shouldered Kite**, **Pied Crow** and **Fiscal Shrike** are common roadside birds along the R27, and they are joined in summer by **Steppe Buzzard** and **Yellow-billed Kite**. From Koeberg northwards, strandveld vegetation is largely dominant and there are a number of places where one can gain access to its birds.

In early 2000, several massive fires swept over much of the West Coast south of Langebaan and devastated huge tracts of its vegetation, which can now be seen in various stages of recovery.

Silwerstroomstrand is the closest prime strandveld site to Cape Town, and supports a rich diversity of birds. Take the 'Silwerstroomstrand' turn-off to the left 32.7 km, north of the 'Dolphin Beach' pans, and park on the edge of the road after 0.8 km.

The thicker vegetation harbours species such as the **White-backed Mousebird**, **Karoo Lark** (common, but inconspicuous), **Cape Penduline Tit** (p.81), **Cape Bulbul**, **Cape Robin**, **Karoo Robin**, **Titbabbler**, **Layard's Titbabbler**, **Grey-backed Cisticola**, **Long-billed Crombec**, **Bar-throated Apalis**, **Grassbird**, **Bokmakierie**, **Lesser Double-collared Sunbird**, **Malachite Sunbird**, **Cape Weaver**, and **Yellow** and **White-throated Canaries**, while the more open patches should be searched for **Grey-wing Francolin** (their calls can be heard early in the morning), **Southern Black Korhaan** (p.57*), **Clapper Lark** (see p.116*; uncommon) and **Cape Bunting**. Even the rare **Hottentot Buttonquail** (see p.23) has been flushed here.

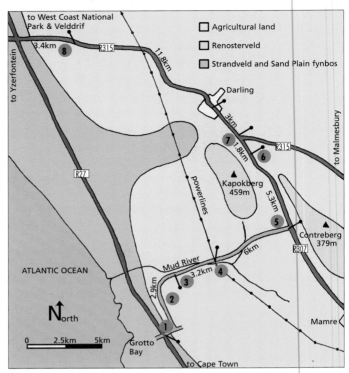

Darling Area and Wildflower Reserves

to West Coast National Park & Velddrif

to Yzerfontein

3.4km

R315

11.8km

Darling

3km

R315

1.8km

to Malmesbury

Kapokberg 459m

powerlines

5.3km

Contreberg 379m

R307

Mud River

6km

3.2km

2.9km

Grotto Bay

to Cape Town

Mamre

ATLANTIC OCEAN

North

0   2.5km   5km

☐ Agricultural land
☐ Renosterveld
☐ Strandveld and Sand Plain fynbos

DARLING FARMLANDS AND WILDFLOWERS

Returning from Silwerstroomstrand to the R27, continue for a further 9.3 km before turning right onto a gravel road marked 'Darling Hills Road' (opposite the conspicuous 'Grotto Bay' sign; look for the yellow flags, ① on site map, above). **Clapper Lark** (p.116*) and **Southern Black Korhaan** (p.57*) occur in the patch of vegetation at the junction of these two roads, and are most conspicuous when they are vocal in spring. Continue along the Darling Hills gravel road for 0.6 km until you reach a small pond (dry in late summer) on the left, where **Avocet** and **Three-banded Plover** can often be seen. **Pied Starlings** have burrowed their breeding tunnels into a sandy bank here, and a male **Pin-tailed**

Roadside birding in the wheatlands of the Swartland, near Darling.

Whydah often displays overhead. This area is most active in winter and spring, when the surrounding wheatlands are filled with birds; after the summer harvest it becomes progressively drier and less active. Small numbers of **Blue Crane** (p.72*) may be found in the adjacent fields (②on map), especially in summer. **Red-capped Lark**, **Capped Wheatear**, **Familiar Chat**, **Grassveld Pipit** and **Cape Sparrow** are common in this vicinity. In spring and summer, look overhead for **Banded Martin** and **Pearl-breasted Swallow** among the more numerous **Greater Striped** and **European Swallows**.

Continue along the Darling Hills road, and look out for the small stream passing under the road at ③. A **Masked Weaver** colony is present in the large tree on your right at this point, and in spring a parasitic **Diederik Cuckoo** lurks around its edges. We have seen all three **mousebird** species perched in a single bush here! The road continues along a winding, overgrown river course until you reach a bridge (④). **White-throated Swallow** breeds under the bridge, and the alien vegetation in this vicinity holds **Fiscal Flycatcher**, **Acacia Pied Barbet** and **Cardinal Woodpecker**. **Titbabbler** and **Long-billed Crombec** may be found in the remnant natural scrub. Listen out at any

stand of exotic trees in this area for **Klaas's Cuckoo**, **African Hoopoe** and the scarce **Greater Honeyguide**. Occasionally, **Secretarybird** is seen stalking the open fields along the road, and **Namaqua Dove** may be seen in summer. The route intersects a tar road (the R307; ⑤on map); just before turning left towards Darling, take a quick scan for the **Jackal Buzzard** that often perches in this area.

If you're passing through these parts at any time between August and October, do pay a visit to the Waylands Wildflower Reserve (⑥) where renosterveld vegetation occurs (see p.7). The vivid colours and massive diversity of flowering bulbs are truly spectacular and are sure to impress even the most hardened of birders. This is also an excellent place to see **Clapper Lark**. In 1998, a major irruption of **Black-headed Canary** (p.105*) and **Ludwig's Bustard** (p.105*), two arid-country species normally found much further north in Namaqualand and the Karoo, occurred in this part of the Cape. Large numbers of both these species invaded the area for some weeks, and the canaries even bred here. Continuing towards Darling, the tiny Oudepos Wildflower Reserve (⑦) provides further great birding and flower viewing, and can be enjoyed from

Male Cape Weaver, a South African endemic.

(summer), **Thick-billed** and **Red-capped Larks**, **Capped Wheatear**, **Orange-throated Longclaw** and **Grassveld Pipit** are also found here. The seasonally flooded marsh often holds **Ethiopian Snipe**, and **Blue Crane** (p.72*) is an occasional early morning visitor to the reserve.

the comfort of your car. You'll see the white gateway to the reserve on your left, directly opposite the R315 turn-off to Malmesbury. A few pairs of **Cloud Cisticola** (p.57*) breed here, and **Orange-throated Longclaw**, **Thick-billed Lark**, **Grassveld Pipit** and **Yellow Canary** are also common. A small group of **Spotted Dikkop**, which can sometimes be quite difficult to see, roosts in the gardens near the main building – which, incidentally, is the largest orchid nursery in the southern hemisphere.

The quaint town of Darling, home to a number of local artists including the internationally renowned political satirist Pieter-Dirk Uys, appears rather unhurried, except perhaps during the peak flower season. It is an excellent place to take a short break from birding and enjoy a light meal. **Little Swifts** breed in the town and can almost always be seen wheeling overhead. **Cape Canary** can also be seen in the town. The Wildflower Reserve, adjacent to the town itself, is worth a visit in spring if you have time.

Continue through Darling along the R315 towards Yzerfontein (poorly signposted), which eventually intersects the R27. The unassuming Tienie Versveld Nature Reserve lies on the left at ⑧. There are no facilities here, but its presence is betrayed by a small sign and a stile over the fence. For most of the year, it resembles an ordinary, abandoned field, but in spring it undergoes a spectacular transformation, becoming a vast mosaic of flowers. It is best known among birders as an excellent site for **Cloud Cisticola** (p.57*), which is common in the areas of taller grass. **Common Quail**

YZERFONTEIN

This is a scenically attractive coastal village that holds a number of quality species. However, it is probably only worth visiting if you are not going any further north. To find **Chestnut-banded Plover**, take the R315 towards Yzerfontein for 4 km from its junction with the R27. Turn right here (signposted 'Gypsum Mine') and continue for a further 1.8 km (the first 1 km of this track is good for strandveld birding) until you get to the mine on the edge of the vast Yzerfontein salt pan. Ask permission at the office before checking the edges of the pan for the plovers. Although water levels fluctuate greatly throughout the year, there are almost always some **Chestnut-banded Plovers** here. Returning to the R315 (keep a look out for **European Bee-eater** here in spring and summer), continue along the main road into Yzerfontein village itself, and search the rocks along the shore for **Crowned Cormorant** and **African Black Oystercatcher** (p.32*). The rocky island beyond the harbour is home to breeding **Bank Cormorant** (see p.21). The nests on the top right of the island belong to this species, and should preferably be viewed through a telescope. Heaviside's Dolphin (see p.104) can often be seen offshore.

WEST COAST NATIONAL PARK

The still, aquamarine waters of the sheltered, 16-km long Langebaan Lagoon, the jewel of the West Coast, provide excellent birding. Granite inselbergs rise sharply from its northern shores, while South Africa's largest saltmarsh lies at its southern end. The West

Coast National Park has become a legendary birding site, best known for the large numbers of migrant waders that crowd the mudflats during summer. These can easily be observed from the well-positioned bird hides, offering local birders an excellent chance of finding rarities. The top-class strandveld birding, spring flowers and proximity to Cape Town (taking the direct route along the R27, it is less than an hour from the city) all make the West Coast National Park a most productive, pleasant, and accessible birding destination.

Approaching from the south along the R27, the well-marked turn-off to the West Coast National Park is 10.9 km beyond the R315 Yzerfontein/Darling junction. An entrance fee, which includes a map and birdlist, is payable at the gate. A meandering tar road leads northwards into the park, passing through some excellent strandveld. Roadside birding in the park is highly rewarding. **Ostrich** are readily seen, resembling giant prehistoric reptilians rather than birds as they stride across the vegetated dunes. **Cape Francolin** is very common throughout the reserve, and coveys of the smaller and scarcer **Grey-wing Francolin** should be carefully searched for on the road edges in the early morning and evening. **Black Harrier** (p.57*) may be seen quartering low over the vegetation anywhere in the park. **Black-shouldered Kite** prefer roadside perches, and many roost communally at night in the large reedbeds on the eastern side of the lagoon, after gathering in one of the lonely palms trees in this area. Flocks of **Pied** and **Wattled Starlings** occur throughout the park. **Southern Black Korhaan** (p.57*) is regularly seen at the roadside, especially between Geelbek and the park's northern exit near Langebaan village.

Because the vegetation is so dense, visitors are unlikely to see many of the mammals that occur here. Two small antelope, Common Duiker (*Sylvicapra grimmia*) and Steenbok (*Raphicerus campestris*), are often startled at the road edges (especially in the early morning), giving a brief view of themselves before darting back into the vegetation. The peculiar tortoise roadsigns along this route refer to Angulate Tortoises (*Chersina angulata*), which are commonly seen crossing the park roads.

The strandveld vegetation throughout the park harbours species such as **White-backed Mousebird, Karoo Lark, Cape Penduline Tit, Cape Bulbul, Cape Robin, Karoo Robin, Titbabbler, Layard's Titbabbler, Grey-backed Cisticola, Long-billed Crombec, Bar-throated Apalis, Grassbird, Bokmakierie, Lesser Double-collared Sunbird, Malachite Sunbird, Cape Weaver, White-throated Canary, Yellow Canary** and **Cape Bunting**.

Check for **Pearl-breasted Swallow** among the flocks of commoner **European** and **White-throated Swallows**.

The Geelbek mudflat bird hide (① on site map overleaf) allows for superb wader watching in summer, and is arguably South Africa's best waterbird hide. The array of desirable vagrant waders that have been found here over the last few years (see p.50) render it the favoured haunt of dedicated twitchers such as Trevor Hardaker, who make the pilgrimage here with fanatical regularity. It allows for close-up views of a large diversity of wading species; common summer migrants include **Curlew Sandpiper, Little Stint, Sanderling, Knot, Turnstone, Greenshank, Marsh Sandpiper** (unusually

Geelbek mudflat bird hide

47

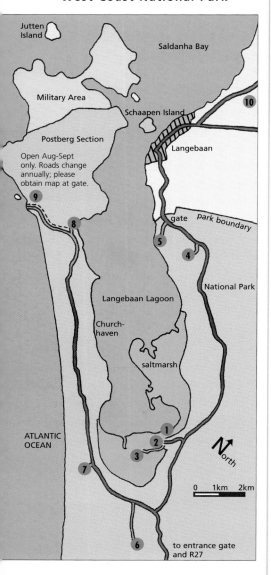

Jutten Island

Saldanha Bay

Military Area

Schaapen Island

10

Postberg Section

Langebaan

Open Aug-Sept only. Roads change annually; please obtain map at gate.

9

gate   park boundary

8

5

4

National Park

Langebaan Lagoon

Church-haven

saltmarsh

ATLANTIC OCEAN

1

2

3

7

North

0   1km   2km

6

to entrance gate and R27

Marsh Sandpiper occurs at ① on map.

the scarce but regular **Terek Sandpiper** and, with luck, **Greater Sandplover** or **Redshank**.

Timing is very important: the area is at its most rewarding on the correct part of the tidal cycle. This is notoriously difficult to predict, although the best viewing usually begins about 4.5 hours after the 'High tide in Table Bay' as listed in Cape Town newspapers. At about this time, the water drops and slowly begins to expose the mud and its invertebrates on the surface. The longer-legged waders land first, soon to be joined by the smaller species as the water recedes further still. If you get the timing wrong, try the Seeberg hide (opposite), which is better at high tide, or console yourself with the view of the distant white cliffs of the lagoon's western shore. These were recently in the international limelight when the most ancient of recorded *Homo sapiens* footprints

LEMMING CYCLES AT LANGEBAAN LAGOON

Professor Les Underhill of the Avian Demography Unit at the University of Cape Town is running long-term water-bird counts at the lagoon, and has demonstrated that the number of **Curlew Sandpipers** wintering here is affected by the Arctic Lemming cycles. When the Lemming numbers in Siberia crash every three years, Arctic Foxes turn to preying on the sandpipers instead.

common here), **Whimbrel**, **Grey Plover**, **Ringed Plover**, **Bar-tailed Godwit**, and a smaller number of resident **White-fronted** and **Chestnut-banded Plovers**. A few of the localized **Curlew** are always present, and it usually takes some careful scanning to pick up were discovered here (having lain preserved in rock for the past 117 000 years).

A wide variety of other waterbirds may be seen from the hide, including **South African Shelduck**. **African Rail** is regularly seen darting in and out of the sedges, especially in the early morning on the right-hand side of the hide. **African Marsh Harrier** breeds in the adjacent reedbeds, and **Osprey** passes overhead in summer. The approach to the hide is by way of a wooden boardwalk that serves to protect a splendid tract of multicoloured saltmarsh. This endangered vegetation type is very sensitive to disturbance and takes many years to recover from damage from trampling. Check the small pools here for **Kittlitz's Plover, Black-winged Stilt, Blacksmith Plover** and **Cape Wagtail**. Noisy **Levaillant's Cisticola, African Sedge Warbler** and **Cape Reed Warbler** occur in the adjacent reedbeds.

The Geelbek manor house (②) on map opposite), restored in the typical Cape Dutch style, has a small restaurant with tame **Cape Francolins** and **Cape Weavers** in attendance. **Acacia Pied Barbet, Tit-babbler** and the occasional **Cardinal Woodpecker** frequent the stands of largely alien trees. A further two bird hides (and another in preparation, sponsored by the Cape Bird Club) are located at a pan on the edge of a vast saltmarsh (③), a 20-minute walk from the manor house. Starting at the parking area, it passes through old farmlands where **Thick-billed Lark, Levaillant's** (near water) and **Grey-backed Cisticolas, Stone-chat** and **Cape Sparrow** are common. The number of birds on this saltpan depends on the water level, and on the state of the tide; high tide is best, as many birds roost here when forced off their mudflat feeding grounds. Mainly smaller waders occur here, including large numbers of **Little Stint**. When water levels are low, **White-fronted, Chestnut-banded** and **Kittlitz's Plovers** are often common. Small groups of **Caspian Plover**, a species unknown elsewhere in the Cape, have in recent years regularly been seen in this vicinity. Keep a look out overhead for **Peregrine Falcon**, which consistently harass the waders.

On the road to Langebaan village, the Seeberg lookout (④) can be seen perched on a granite hillock, and provides a panoramic view of the lagoon. It is worth scanning for **Black Harrier** (p.57*) in this vicinity. The granite boulders below are home to a group of dassies (Rock Hyrax *Procavia capensis*). The viewsite is reached by a short, unsurfaced road that is probably the best place in the park to look for **Southern Black Korhaan** (p.57*). This small bustard is often seen at close quarters in the open areas near the viewsite parking area, especially in the morning and evening. Also look out for **Grey-wing Francolin**.

The short road down to the Seeberg hide (⑤) offers good strandveld birding, and is covered with flowers during spring. The low ridge of dunes on the right, about 100 m down the road, is home to a covey of **Grey-wing Francolin**, which are almost always found in this vicinity. Most of the strandveld species can be seen at the parking area near the hide, including **Cape Penduline Tit** (p.81) and **Layard's Titbabbler**. Although

Cape Francolin are common in the park.

the Seeberg hide is not as well situated as that at Geelbek, it offers a good selection of waders and terns, especially at high tide. Large numbers of **Bar-tailed Godwit** may be seen here, and **Little Tern** often roosts on the closest sandbank.

Abrahamskraal waterhole (⑥) is one of the only sources of fresh water in the park, and many birds come here to drink, including **Namaqua Dove**, **Wattled Starling**, **White-throated** and **Yellow Canaries**, and **Cape Bunting**. **Cape Reed Warbler**, **African Sedge Warbler** and **Levaillant's Cisticola** are common and easily seen in the reeds, while **Black Crake** skulks lower down at the reed bases. A variety of widespread waterbirds also occur here, including **African Spoonbill**, **Moorhen** and **Three-banded Plover**. **Brown-throated Martin** hunts overhead.

The viewsite at ⑦ offers not only panoramic views over the lagoon and sea, but is a good spot to look for **Karoo Lark**. The northwestern Postberg section of the park (⑧) is open only during the flower season (August to October). Dominated by sloping meadows strewn with granite boulders, this part of the reserve offers spectacular scenery, excellent flower viewing, pleasant birding and a variety of introduced game species, including Gemsbok (*Oryx gazella*) and Springbok (*Antidorcas marsupialis*). There is year-round

FINDING RARITIES

Summer, spring and autumn are the best times to search for rare waders, although there is also a peculiar early winter peak of 'reverse migrants' – birds wintering in central Africa which inadvertently migrate south to the Cape instead of north to Europe. Winter is also the best time for errant **American Purple Gallinules** (search reedbeds on the Peninsula and western seaboard) and **Greater Sheathbills** (most often seen on the Peninsula's rocky coasts, notably the Atlantic seaboard). Rare waders have turned up all over the region, although the top two sites are undoubtedly Langebaan Lagoon in the West Coast National Park (especially the Geelbek mudflat and saltmarsh hides, p.47) and the Berg River estuary (especially the Riviera and De Plaat mudflats and Cerebos salt works, p.53). Other notable sites include Wadrif Saltpan (p.55) and, on the Cape Peninsula, the Strandfontein sewage works (p.26) and the rocky shores of the Cape of Good Hope Reserve, near Olifantsbos (p.24). Please see p.11 for details of how to report any unusual sightings.

Exciting rarities, including Little Blue Heron and Lesser Yellowlegs, have been viewed from the Riviera mudflat bird hide at Velddrif on the Berg River estuary (⑦ on map on p.53).

Seasonality of rare birds recorded in western South Africa

Africa's most southwestern corner is known throughout southern Africa as the region's rarity hotspot (see box opposite). While many of these vagrants are pelagic seabirds (see *Seabirding*), the majority are migrant waders from Europe and the Americas. There is seasonal variation between species, and all the western South African records of important vagrants are summarized in the table below.

	JANUARY	FEBRUARY	MARCH	APRIL	MAY	JUNE	JULY	AUGUST	SEPTEMBER	OCTOBER	NOVEMBER	DECEMBER	TOTAL NO. RECORDS
European Oystercatcher			1			2	2	1	1	1		1	6
Caspian Plover	1	1	1									2	5
American Golden Plover	2	2	2						1			2	8
Pacific Golden Plover			2										2
Green Sandpiper	1											1	2
Redshank	2	4	2	2	1		2	2	1	1	2	3	18
Lesser Yellowlegs	2	2	1	1	1	1	1	1	1	1	1	2	2
Greater Yellowlegs											1		1
Dunlin	2	1										1	2
Red-necked Stint	1		2	2				1	1	1	1	1	7
White-rumped Sandpiper										2	1	1	4
Baird's Sandpiper										1			1
Pectoral Sandpiper	3	3	5	1						2	1	2	10
Buff-breasted Sandpiper									1				1
Broad-billed Sandpiper	3	5	1						1			2	9
Black-tailed Godwit	1	1						2	1	1	1	2	5
Hudsonian Godwit	2	3	1			1			2			2	5
Grey Phalarope	1	1	1							2	1	1	4
Red-necked Phalarope	2	3	2							1	1	1	7
Wilson's Phalarope	2	1	2				1			1	2		7
Crab Plover				1									1
American Purple Gallinule					4	2	4			1			9
Greater Sheathbill	1				1	2	4	4	1	3		1	14

All records included in this list are those accepted by the South African National Rare Bird Committee, with a few exceptions having been made for recent, well-substantiated records. This table should be treated as a guide and not a reference.

access to the sea at Tsaarsbank (at ⑨, outside the Postberg section), where **African Black Oystercatcher** (p.32*) and **Cape, White-breasted** and **Crowned Cormorants** occur on the rocks. **Cape Gannet** and **White-chinned Petrel** can often be seen offshore.

LANGEBAAN QUARRY

The greatest attraction here is the resident pair of **Black Eagles**, which usually breed between May and November, although they may be seen in the general vicinity throughout the year. At the northern edge of Langebaan village, turn left towards Club Mykonos. After a few kilometres along this road, turn sharply to the right opposite the horseshoe-shaped 'Long Acres' sign. Follow the main track for about half a kilometre, veering right wherever it splits (⑩ on map, p.48). This will take you to the section of quarry where the well-known eagles' nest is situated (it is just right of centre on the main cliff face, but remarkably inconspicuous despite its size). **Rock Kestrel** and **African Black Swift** also breed on the cliff faces. The alien thicket at the quarry edges is a reliable site for **Southern Grey Tit** and **Acacia Pied Barbet**.

VREDENBURG TO PATERNOSTER AND ST HELENA BAY

The granite outcrops, agricultural lands and scrubby vegetation of the Columbine Peninsula provide access to a suite of species not easily available elsewhere in this region. Vredenburg, the administrative centre of the region, can be conveniently reached from the R27, or from Langebaan. Take the Paternoster road west of Vredenburg, and scan along the fence posts for **Sickle-winged Chat** (remarkably common), **Anteating Chat** and **Yellow Canary**. Regularly seen roadside raptors include **Jackal Buzzard** and **Lanner Falcon**. Continue westwards to the picturesque fishing village of Paternoster, before heading north towards Stompneusbaai and St Helena Bay. Check the fallow lands 2 km along this road for **Cape Long-billed Lark**, as well as **Thick-billed Lark**, **Red-capped Lark** and **Grey-backed Finchlark** (the latter especially during late summer). Small patches of natural vegetation harbour the regular strandveld species in addition to the scarcer **Southern Grey Tit**, **Cape Penduline Tit** (p.81) and, rarely, **Yellow-bellied Eremomela**. Check small patches of water for **South African Shelduck**, and the occasional drinking **Namaqua Sandgrouse**. From Stompneusbaai, the road runs along the rocky shoreline of St Helena Bay (look for roosting cormorants and **African Black Oystercatcher**) before reaching the Berg River estuary at Velddrif.

THE BERG RIVER ESTUARY, VELDDRIF

The estuary and floodplain cover a vast area, extending 40 km inland along one of the Cape's biggest rivers. This area encompasses a wide diversity of habitats, including sandy beaches, mudflats, reedbeds, riverine channels, strandveld and floodplain, and it is famously rich in birdlife. The floodplain itself, which is very seasonal and difficult to access, holds few birds that cannot be seen nearer the mouth. Rather, it is the mudflats and saltpans that provide the most rewarding birding, and have proven excellent for rarities (see box, p.50). Ornithologists at the Percy FitzPatrick Institute (see p.57) have studied this region intensively and have shown that these mudflats support the highest density of waders along the entire east Atlantic seaboard.

Chestnut-banded Plover: Cerebos salt works.

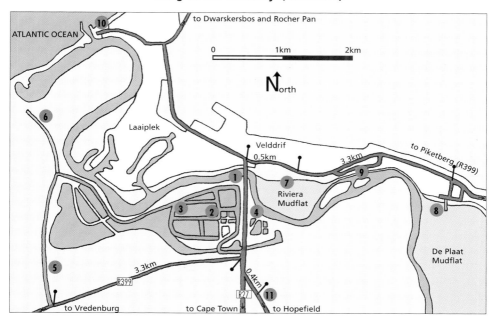

After 140 rather monotonous kilometres from Cape Town, the R27 coastal road spans the Berg River and enters the town of Velddrif, which sprawls along its northern bank. Just before the road crosses the river (① on map above), the Cerebos salt works (②) lies to the left. The evaporation pans of this conveniently non-tidal locality are particularly good for **Chestnut-banded Plover**, and roosting waders that flee the mudflats at high tide. Another advantage here is that the birds are often closely approachable from the comfort of your vehicle; ask permission at the office to drive around (but don't try this in winter when the rains make the roads extremely slippery).

The pans closest to the offices (② on map) are the most reliable for **Chestnut-banded Plover**, which may be seen here in fair numbers. The bulk of South Africa's population of this localized species occurs on the saline pans of the West Coast. **Kittlitz's Plover** is also common here, and often nests on the edges

White Pelicans at the Berg River estuary.

of the roads that skirt the pans. During summer, large numbers of waders feed along the pan margins; the most common species include **Little Stint**, **Ruff**, **Ringed Plover** and **Curlew Sandpiper**. Other birds here, especially in the vicinity of ③, include **Black-necked Grebe**, **Greater** and **Lesser Flamingo**, **Cape Teal** and **Caspian Tern**. It is also worth scanning the pans on the east

53

Cape Cormorants nesting on fishing boats.

side of the R27 at ④, which hold a selection of small waders, **Black-winged Stilt**, **Avocet** and **South African Shelduck**. **Red-necked Phalarope** is recorded here infrequently (but beware of distant **Ruff** that often feed by spinning in circles in an alarmingly phalarope-like manner!). Also look out for **Thick-billed Lark** and **Capped Wheatear** in the salt works.

To reach the back entrance of the salt works, take the R399 Vredenburg road and follow the Flaminkvlei right turn ⑤. This road initially passes some farmland: **Thick-billed Lark** and **Stonechat** occur here, and **Jackal Buzzard** is often seen in this vicinity (but note that the juveniles can resemble **Steppe Buzzard**). **Greater Flamingo** is often observed in this part of the salt works. Just beyond the buildings, a sandy track leads leftwards towards the coast and the strandveld along here ⑥ holds **Cape Long-billed Lark**.

The Riviera mudflat is perhaps the most famous of the Velddrif birding localities, and a number of rarities have been seen here – most famously the only African record of the American **Little Blue Heron**. The bird hide is situated at ⑦ (ask at the adjacent filling station or restaurant for the key). As always, it is important to get the tide right: the best viewing begins about 1.5 hours after the 'High tide in Table Bay' as listed in Cape Town newspapers. A diverse selection of waterbirds may be observed from here, including **White Pelican**, **African Spoonbill**, **Greater** and **Lesser Flamingo**, **Purple Heron**, **Little Egret**, and a wide selection of terns including **Caspian** and **Little Terns**. The mudflats are excellent in summer for migrant waders, and the selection of species is very similar to that at the Geelbek mudflat hide (refer to p.47). Confiding **Levaillant's Cisticola** call from the nearby sedges.

The De Plaat mudflat ⑧ holds similar birds (and **Curlew** and **Bar-tailed Godwit** are easier to see here), although it is not as easy to approach them closely. De Plaat can be reached by taking the R399 and turning right into Vrede Road (opposite the 'Spreewal Kafee' building). Turn right again, follow this road to its end, and walk down past the eucalyptus trees to the wooden jetty. Again, be sure to visit on the correct tide: the best viewing begins about 3.5 hours after the 'High tide in Table Bay'.

To reach an area of riverine channels and reedbeds (⑨ on map), where a variety of herons, warblers and other waterbirds may be found, take the main road east past the bird hide, and turn right along the road signposted 'Bokkoms Industry' (referring to a West Coast speciality: malodorous dried fish) and continue to the banks of the river. Wooded residential areas of Velddrif, such as those in this vicinity, usually provide **Red-faced Mousebird** and **Acacia Pied Barbet**.

At the harbour on the river mouth in Laaiplek (⑩), a selection of cormorants, gulls (a good site for the scarce **Grey-headed Gull**) and terns roost. The latter are harried in summer by **Arctic Skua** just offshore.

A pair of **Spotted Eagle Owl** is resident at the copse of eucalyptus and wattle trees at ⑪, along the Hopefield road. An excellent area for strandveld birds may be reached along the R27, 10 km south of Velddrif, opposite the turnoff to 'St Helena/Stompneus'. Continue along the unsurfaced road to the east for 1 km, and bird the strandveld near the crest of the hill. Birds present here include **Southern Black Korhaan** (p.57*), **White-backed Mousebird**, **Karoo Lark**, **Southern Grey Tit**, **Cape Penduline Tit**, **Layard's Titbabbler**, **Long-billed Crombec**, **Bar-throated Apalis** and **White-throated**

Canary. **Cape Long-billed Lark** (p.13) is occasionally seen near the R27 roadside, just to the south of here, although beware of confusion with the much more common **Thick-billed Lark**.

The ephemeral Rocher Pan, a large waterbody protected in a nature reserve 24 km north of Velddrif, may be reached via the hamlet of Dwarskersbos. It often supports an interesting selection of waterbirds, and is flanked by very productive strandveld vegetation and a lonely coastline. A desolate beach stretches towards the seemingly limitless horizon, populated only by the occasional pair of **African Black Oystercatchers**.

LAMBERT'S BAY TO VERLORENVLEI

The **Cape Gannet** colony at Lambert's Bay is a spectacle not to be missed, and must rank as one of the birding highlights of the West Coast. Nearly 14 000 pairs breed on the bay's Bird Island, now connected to the mainland by a wide concrete breakwater extending from the harbour (see box opposite). Small numbers of **African Penguin** (p.32*) can also be seen here, and all four marine cormorants breed on the island. A host of gulls and terns, including **Swift Tern**, are also present. Cape Fur Seals (*Arctocephalus pusillus*) may also be seen in the vicinity.

A good selection of waterfowl and waders usually inhabit Jakkalsvlei, a lake on the northern edge of town (reached from the caravan park). Note, however, that it can be dry for the most part in summer. Regular species here are **Greater Flamingo**, **South African Shelduck**, and **Cape** and **Red-billed Teals**. The handsome and localized Heaviside's Dolphin (see box, p.104) sometimes comes close inshore. The strandveld vegetation near Lambert's Bay holds all the birds profiled on p.44; notable strandveld birds include **Clapper** and **Karoo Larks**, **Pearl-breasted Swallow**, **Yellow-bellied Eremomela** and **Rufous-eared Warbler**.

Lambert's Bay can be reached most efficiently by following the N7 national road from Cape Town to Clanwilliam, and then taking the tarred R364 towards the coast (look out for **Anteating Chat** along the way). It can thus be easily visited as a detour from the N7 while en route to Namaqualand or Bushmanland, and is well combined with a visit to Kransvlei Poort (see overleaf). Those with a little more time may wish to travel along the rather poor unsurfaced roads leading south from Lambert's Bay towards Velddrif, that offer pleasant wetland and strandveld birding (see below).

Follow the coastal road south from Lambert's Bay towards Elands Bay (look out for the uncommon **Chat Flycatcher**), and, after 11.8 km (just before the railway bridge), turn right to follow the railway line until you reach Wadrifsoutpan ('wagon drift saltpan') after about 1 km. This is a private road, and you are not permitted to proceed

---

CAPE GANNET COLONY

This vibrant colony is the most accessible of the six **Cape Gannet** colonies in existence. Huge numbers of gannets are present throughout the year, and can be viewed face to face from a modern, newly constructed hide. A panel of one-way glass forms one side of the hide, allowing undisturbed observation of the gannets' curious behavioural interactions. Amazingly, a few individuals of the vagrant **Australian Gannet** annually linger in an offshore Cape Gannet colony further south.

past the toll station adjacent to the pan. Wadrifsoutpan is split in two by the railway line, and the smaller seaward section is worth searching for a selection of waterbird and wader species, including **South African Shelduck**, **Cape Teal** and **Greater Flamingo**. However, it can be largely dry in summer. A wide variety of strandveld birds occur here, most notably **Cape Long-billed** and **Clapper Larks** (p. 116*).

Continue along the unsurfaced road to Elands Bay, which is situated at the mouth of the bird-rich Verlorenvlei ('lost lake'). At Elands Bay, turn southwards along the road that crosses the vlei, and turn to the left at the T-junction on the southern bank. Scan the reedbed edges (such as those in the vicinity of the road bridge) for **Little Bittern**, **African Rail**, **Red-chested Flufftail**, **Purple Gallinule**, **Purple Heron**, **Malachite Kingfisher** and **African Marsh Harrier**. The rocky slopes lying south of the T-junction hold a host of scrub birds, including **Southern Grey Tit**. A pair of **Black Eagles** breeds on the nearby cliffs and are often seen overhead. You may wish to continue along the southern edge of the lake for a few more kilometres, as a wide diversity of waterbird species may be seen from the road. These include **Great Crested Grebe**, **White**

Pelican, **Greater** and **Lesser Flamingoes**, **South African Shelduck**, **African Fish Eagle**, **Caspian Tern** and a variety of waders.

Retrace your route to Elands Bay, then turn right onto the R366. This follows the 40-km length of Verlorenvlei inland towards Redelinghuys (please ask the landowners' permission should you wish to reach the lake itself at any point), before heading south to Aurora. The mountains to the east of Aurora hold a number of interesting species, including **Protea Canary** (opposite), and **Black** and **Booted Eagles**. The tarred road resumes from Aurora southwards, and ultimately intersects with the R399, 41 km to the east of Velddrif.

KRANSVLEI POORT

Tucked away close to the N7 national road, just over 200 km north of Cape Town, Kransvlei Poort is a highly accessible and reliable site to see the often-elusive **Protea Canary** (see opposite). From the N7, exactly 10 km south of Clanwilliam, turn left onto the gravel road marked 'Paleisheuwel'. Just over 2 km further on, the road enters the 'poort' (see box, p.8). Here the road follows a reed-lined stream through a steep valley lined with low cliffs.

Search for **Protea Canary** in the taller vegetation fringing the roadside and the lower cliffs, especially near the prominent bend in the road towards the end of the poort. The protea stands that line the route as it rises out of the poort are, oddly, not the best place to look for this bird. Other canaries found in the poort include **Streaky-headed**, **White-throated**, **Bully** and **Cape Canaries**. **African Sedge Warbler** and **Yellow-rumped Widow** are found along the stream, and **Ground Woodpecker** (p.105*) frequent the cliffs and rocky outcrops. **Layard's Titbabbler**, **Long-billed Crombec** and **Fairy Flycatcher** occur in the hillside scrub. **Cape Eagle Owl** (p.105*) occurs in this vicinity. Visit this site en route to Lambert's Bay, Namaqualand or Bushmanland.

Kransvlei Poort lies on the edge of the Cederberg Mountains.

SELECT SPECIALS

Black Harrier

This striking harrier is one of four raptor species endemic to southern Africa. It ranges widely over scrub and grassland in western South Africa, and is most regularly encountered in the West Coast National Park. While the pied adults are very distinctive, immature birds regularly pose identification challenges. They are best recognized by their combination of a white rump, white undersides to the inner flight feathers (which result in a pale patch on the underwing), dark upperparts and brown, streaked underparts. The Black Harrier is the emblem of the Percy FitzPatrick Institute of African Ornithology, an organization of international repute based at the University of Cape Town and involved in research on the ecology, evolution and conservation of the continent's birds.

Southern Black Korhaan

This small bustard, endemic to the winter-rainfall areas of South Africa, is one of the most characteristic species of the West Coast and is held in fond regard by the locals. It is sexually dimorphic, and the strikingly plumaged males produce a raucous, grating call in spring. It has been split from the Northern Black Korhaan of the interior grasslands on the basis of differences in call, display, size, plumage and examination of genetic material (see p.12). Although it is widespread throughout the region and may be

seen anywhere, it is best found by searching the road edges in the West Coast National Park (p.49). Korhaan is an Afrikaans word that refers to small bustard species, and is derived from the Dutch word for the Palaearctic Black Grouse (*Tetrao tetrix*).

Protea Canary

Protea Canary is regarded as one of the most elusive of the fynbos endemics, largely because it is uncommon close to Cape Town. It can, however, be quite common in many of the less accessible mountainous areas of the region, such as the Cederberg Wilderness Area. For visitors without the time to venture so far off the beaten track, the best areas to search for it are Kransvlei Poort (p.56), Paarl Mountain (p.82), Mitchell's Pass (p.81), and, further afield, Swartberg Pass (p.123). Although inconspicuous, it draws attention to itself by its distinctive song. It is by no means restricted to protea stands; in fact in many areas it appears more common in tall, non-protea vegetation.

Cloud Cisticola

The Cloud Cisticola is best detected in spring when the calling males are visible as distant, almost imperceptible specks fluttering high in the air during their undulating display flight. Good views can often be obtained by waiting patiently until they eventually drop sharply to land in the grass. The southern Cape subspecies is distinct from others further north, in South Africa and Zambia, both vocally and by its conspicuously streaked breast, which is an excellent field character to separate it from the otherwise dauntingly similar Fan-tailed Cisticola. There are several indications that this distinctive subspecies may be a full species (see p.13). It is best found in grassy and agricultural lands, especially the Tienie Versveld and Oudepos Wildflower Reserves (pp.46 and 45) and the Overberg wheatlands of the south coast (p.64).

The Overberg and South Coast

'The mountains rise nobly at a couple of miles distance or so, their bases were lost in a bluish vapour – greenish hillocks rose between us and them – 'tis between them and the mountain, in a glen, the woods are to be found which have been reckoned so luxuriant'. LADY ANNE BARNARD, 18TH-CENTURY UNOFFICIAL FIRST LADY OF THE CAPE: DIARY ENTRY ON THE SWELLENDAM AREA, 1798

Across the sandy, low-lying flats that lie east of Cape Town, a barrier of mountains interrupts the landscape. These are the Hottentots Holland, so named by early Dutch settlers who considered them the 'homeland' of the indigenous Khoikhoi ('men of men') peoples, then known as Hottentots. On the far side of these mountains, between the Langeberg range and the ocean, is the fertile Overberg, a gently undulating coastal plain that today lies predominantly under wheat. This region provides a large diversity of much-coveted species, from Cape Rockjumper to Blue Crane.

The Hottentots Holland mountains border the western Overberg and are traversed, via Sir Lowry's Pass, by the N2 national road. The pass is legendary in birding circles for the numerous fynbos specials that are easily accessible just a short walk from the highway. The site is close to Cape Town and can easily be tackled in a morning out of the city. Just to the south is a spectacular coastal drive winding along the eastern coast of False Bay

TOP BIRDS
Cape Vulture, Blue Crane, Stanley's Bustard, Damara Tern, Narina Trogon, Knysna Woodpecker, Agulhas Long-billed Lark, Clapper Lark, Cape Rockjumper, Victorin's Warbler, Southern Tchagra.

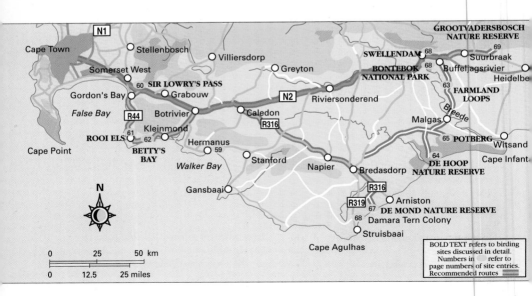

De Hoop Nature Reserve offers a wide variety of habitats, and includes a breeding colony of Cape Vultures (below).

and through the coastal villages of Rooi Els and Betty's Bay, where further fynbos and coastal specials are on offer.

Although all the sites described in this chapter are manageable in a long day trip out of Cape Town, a more extensive two- to three-day loop would be more relaxing and productive for those with the time to spare. A good three-day loop might begin at Sir Lowry's Pass before continuing east, along the N2, to the town of Swellendam, at the foot of the Langeberg mountain range. From here, you can make an eastwards loop to Grootvadersbosch, the Langeberg's largest afromontane forest, for the best diversity of forest birds reasonably close to Cape Town. Returning to Swellendam, where a good variety of accommodation is available, you can strike south through the wheatlands in search of the localized and recently split **Agulhas Long-billed Lark** (see p.73). The agricultural lands also offer numerous other sought-after species, which are surprisingly accessible in this highly transformed land-scape. Nearing the coast of the Indian Ocean, one can enjoy some excellent birding at Potberg mountain in the De Hoop Nature Reserve, before perhaps continuing south-westwards to De Mond, just east of the rather anticlimactic southernmost point of the African continent, at Cape Agulhas. De Mond is noted for its breeding colony of **Damara Tern** (p.68), a highly threatened,

diminutive and attractive species endemic to the South African and Namibian coasts.

Returning towards Cape Town, you might consider visiting Cape Agulhas and, further west, the resort town and harbour of Hermanus. This is one of the most famous whale-watching localities on earth and home to the world's only whale-crier, who wields a kelp-horn to inform one of the whales' appearances. Southern Right Whales (*Eubalaena australis*) are the commonest species, and from July to November you can be sure to see impressive numbers especially close inshore. On your way back to Cape Town, you may wish to visit Betty's Bay and Harold Porter before setting off on the scenic drive that twists along the coast to rejoin the N2 national road at Somerset West.

SIR LOWRY'S PASS

This is a classic Cape birding spot in the Hottentots Holland mountains, and provides easy access to two of the fynbos endemics (**Cape Rockjumper** and **Victorin's Warbler**) that, puzzlingly, do not occur on the Cape Peninsula, despite an abundance of apparently ideal habitat. Excellent fynbos birding may be had minutes from the viewsite next to the N2 highway at the summit, just 50 minutes' drive from the city (along the way, look out for Cape Town's steadily increasing **House Crow** population in the vicinity of the N2 airport off-ramp). A rewarding birding walk at Sir Lowry's Pass can be completed in just two to three hours – longer if you're waylaid by the remarkable plant diversity of these mountain slopes.

The first hurdle lying between visiting birders and their quarry is a blind corner, on the N2 highway: this needs to be crossed on foot with considerable caution after parking at the viewsite on the southern side of the road (at ① on map, opposite). Running north of the road is a rocky ridge of minor outcrops leading up to the summit of Kanonkop peak at ②. Winding along its eastern contour is a broad track ③ leading north towards a neck in the mountains at ④ called Gantouw Pass (after the Khoikhoi

word for Eland, as this was once the route taken by migrating antelope). Here you will see deep ruts in the soft sandstone, a legacy of the east-bound ox-wagons of traders and those who became restive under British rule in the Cape two centuries ago. By 1821, 4 500 wagons a year were making the crossing, a journey of such epic proportions that one in five wagons never survived it! Close by lie a pair of antique signal cannons that were later installed at the pass. The rocky slopes here are the domain of the **Cape Rockjumper** (p.73*). The entire length of the ridge between the N2 viewsite and

FINDING CAPE ROCKJUMPER

The most accessible area to search for rockjumpers (see also p.73*) is the rock-strewn slope at ④ on site map, to the south of Gantouw Pass. To reach this slope, follow the gravel track until it intersects with some powerlines, and turn left onto a small, inconspicuously marked footpath leading up to Gantouw Pass. Walk up to the signal cannons, and work your way to the left (southwards) up the slope, keeping an eye out for rapidly scurrying silhouettes on the clusters of boulders. Note that searching for rockjumpers typically involves scrambling along rocky outcrops and is only recommended to visitors confident of their agility! You can also gain access to the ridge at its southern end, just north of the N2. An inconspicuous footpath at ⑤ leads up the slope from the gravel track. At the ridge, the path peters out and you will need to work your way northwards along the series of outcrops. Cape Rockjumpers are to be found here, but we must stress that this is not an easy walk.

Looking west from Sir Lowry's Pass over False Bay.

the summit of Kanonkop is in fact prime **Cape Rock-jumper** country, and birders alert to its loud, piping call can be sure to locate a group of these fine birds here (see box opposite).

Birds are scarce in this landscape, but the area does have its rewards. The series of rocky outcrops along the path and the ridges above also hold low densities of **Ground Woodpecker** (p.105*), **Familiar Chat, Cape Siskin** (p.33*), **Cape Rock Thrush** and, rarely, **Sentinel Rock Thrush**. Common birds of the dense fynbos between the ridge and gravel track are **Grassbird, Orange-breasted Sunbird** (p.33*), **Neddicky**, and **Karoo Prinia**. **Cape Sugarbird** (p.33*) and **Yellow-rumped Widow** occur more sparsely in denser vegetation, such as that growing along the stream under the powerlines.

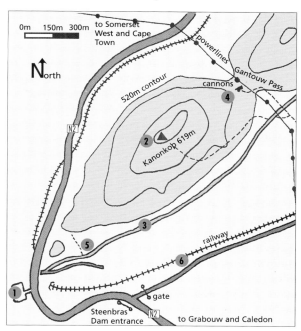

Sir Lowry's Pass

Victorin's Warblers (p.73*) can be heard singing from the slightly denser vegetation of the hill slopes. Those unfamiliar with its call should take care to distinguish it from the superficially similar, but less repetitive, song of the more conspicuous **Grassbird**. Victorin's Warblers are far more readily lured from cover in this relatively open habitat than in their more typical haunts in impenetrable streamside thicket, such as that found beneath the powerlines at Gantouw Pass, where they are common. Another good area to look for them is in the dense vegetation along the railway track at the bottom of the mountain at ⑥. To reach this spot, drive past the viewsite, turn left opposite the entrance to Steenbras Dam, and park at the gate. This area also supports **Striped Flufftail**, although this species is almost impossible to see during the day.

Jackal Buzzard (including at least one potentially confusing white-breasted individual; see p.99), **Rock Kestrel** and **Peregrine Falcon** are the most frequently seen raptor species at Sir Lowry's Pass, although numerous others occasionally pass through the area. These include **Black** and **Martial Eagles, Black Harrier** (p.57*) and **Red-breasted Sparrowhawk**.

Over the crest of Sir Lowry's Pass, the N2 continues eastwards across a rolling plateau, largely covered in timber plantations and South Africa's most important deciduous fruit orchards, before dropping suddenly into the wheat-swathed lowlands of the Overberg (see p.63).

ROOI ELS TO BETTY'S BAY

East of False Bay, the contours of the Hottentots Holland plunge precipitously into the sea, creating a spectacular stretch of coastline where mountain fynbos and marine specials can be seen virtually alongside each

Disa Kloof in Harold Porter Botanical Garden.

other. Heading east of Cape Town on the N2 national road, pass through Somerset West and turn right onto the R44 (signposted 'Gordon's Bay/Kleinmond'). Continue to the T-junction at the edge of the town of Gordon's Bay, and turn left onto the scenic and dramatically sinuous coastal road that meanders southwards to the holiday village of Rooi Els.

At the edge of the village, the road crosses the Rooi Els river before veering to the left and heading steeply up a hill towards Betty's Bay. To look for **Cape Rockjumper** (p.73*), take the second turn-off to the right (an unsurfaced road), just as the R44 begins its ascent. Park at the gate about 1 km further on, and continue on foot. Look out for rock-jumpers on the left-hand side for the next 2 km, and for **Black Eagle**, **Ground Woodpecker** (p.105*) and **Cape Rock Thrush**.

A little further east on the R44 lies the village of Betty's Bay and nearby Stoney Point, site of one of only two mainland colonies of **African Penguin** (see pp.24 and 32*). On the lower mountain slopes of the dramatic Kogelberg range is the Harold Porter Botanical Garden, offering good fynbos and forest birding. To reach Stoney Point, take the signposted right turn towards the coast, just before the series of lakes on your right as you enter the rambling village of Betty's Bay. The **African Penguin** colony is smaller than that at Boulders Beach on the Cape Peninsula (p.24) and usually hosts less than 100 breeding pairs. One unique problem that has faced the Stoney

Point colony was devastating predation by a Leopard (*Panthera pardus*) that descended from the adjacent mountains. **Bank Cormorant** (p.21) also breeds here, alongside the more common **Crowned, Cape** and **White-breasted Cormorants**.

The Harold Porter Botanical Garden lies conspicuously signposted on your left just past the commercial centre of Betty's Bay, and is a fine place for a few hours gentle ramble. The cultivated gardens are quite small, but surrounded by moist mountain fynbos dissected by two forested ravines, Disa Kloof (to your left) and Leopard Kloof (to your right). From the entrance gate, head up through the cultivated gardens to the solidly-built bridge over the Disa Kloof stream, and continue along the path up the kloof itself. Common birds of the lower gardens are **Black Saw-wing Swallow, Cape Bulbul, Karoo Prinia, Southern Boubou, Malachite, Orange-breasted** (p.33*) **and Lesser Double-collared Sunbirds, Yellow-rumped Widow, Bully Canary** and conveniently accessible **Cape Siskin** (p.33*). **Protea Canary** (p.57*) occasionally wanders down into the gardens, but is by no means regular here. There are usually swifts and swallows foraging overhead (including **Rock Martin** and **African Black** and **Alpine Swifts**), alongside soaring raptors (most commonly **Black Eagle** and **Jackal Buzzard**).

The forested path up Disa Kloof leads to a small, bitterly black dam, and ends a few hundred metres further on where a waterfall interrupts the stream. The forest along this path provides **Bar-throated Apalis, Cape Batis, Paradise** (summer) and **Dusky Flycatchers**, and occasionally **Swee Waxbill**. During late summer, spectacular Red Disa (*Disa uniflora*) orchids can be seen clinging to the dripping and slippery cliffs adjoining the waterfall. Make your way back to the dam and cross the bridge over its wall. A pair of **African Black Duck** is often present on the dam, if not elsewhere along the stream. A gentle path then leads out of the kloof and

Blue Cranes dance in the Overberg farmlands, with the Langeberg Mountains in the background.

around the buttress between Disa and Leopard Kloofs, before dropping back down into the gardens. **Cape Siskin, Neddicky, Orange-breasted Sunbird, Victorin's Warbler** (p.73*), **Grassbird** and **Ground Woodpecker** occur along this path. To visit Leopard Kloof, ask for a key at the entrance to the gardens. The forest conceals a series of pleasant waterfalls and, in addition to the forest species mentioned above, also hosts **Olive Woodpeckers** and the Cape's western-most regularly occurring **Blue-mantled Flycatchers**, which sometimes wander down into the cultivated gardens.

    Cape Eagle Owl (p.105*) may be seen in the gardens at dusk by only the most fortunate of birders, and has even been observed at the roadside in Betty's Bay village. Listen for its deep call, especially in winter.

The black water lakes and seeps around Betty's Bay hold an exciting diversity of amphibians, including the Micro Frog (*Microbatrachella capensis*), one of the world's most threatened frog species (pictured left).

FARMLAND LOOPS

The superficially sterile monoculture of the Overberg wheatlands harbours a surprising diversity of birds, including such sought-after species as **Black Harrier** (p.57*), **Blue Crane** (p.72*), **Stanley's Bustard** (p.72*), **Karoo Korhaan, Agulhas Long-billed Lark** (p.73*) and the endemic southern Cape subspecies of **Clapper Lark** (see pp.64, 116*). The area is also pleasantly scenic, with only the scatter of fiery red aloes across the winter hillsides destroying the illusion of a restful southern European landscape.

    One of the best birding areas to explore is that between Swellendam and De Hoop Nature Reserve. Three good gravel roads (see map, p.58) run between the two, flanked by a mosaic of wheatfields, fallow lands, and, on the steeper hillsides and valleys, islands of natural renosterveld scrub (see p.7). A rewarding loop that offers access to all the important birds is the following: take the N2 national road past Swellendam, and continue for 7 km to the hamlet of Buffeljagsrivier. Just beyond the BP service station, turn right onto the gravel road (sign-posted 'Malgas'); turn left after 3.3 km and continue for a further 4.3 km before pulling off. Search the scrub along the road edge for **Agulhas Long-billed Lark** and **Clapper**

Agulhas Long-billed and Clapper Larks can be found in the roadside scrub.

The marjoriae subspecies of Clapper Lark occurs in the Overberg. Below: Pied Starling

Lark. Both are common here and are especially conspicuous when aerially displaying in spring. This road is also good for the scarce **Stanley's Bustard**, **Karoo Korhaan** (rather atypically, in such moist habitat), **Southern Black Korhaan** (p.57*), **Greywing Francolin** (seen feeding on the road verges in the mornings and late afternoons) and **Long-billed Pipit**. Exactly 28.3 km from the N2, shortly after you cross two cattle grids, look for one of the Cape's few **Horus Swift** colonies in a gully on the western (right-hand) side of the road. Four kilometres further on, the road crosses the Breede River at the village of Malgas. Here, you can enjoy the quaint experience of having your car inched across the river on South Africa's last working pont.

Just past Malgas, the route joins the gravel road that leads to Potberg and the De Hoop Nature Reserve, and ultimately to the town of Bredasdorp. The remnant patches of indigenous scrub near this junction are good for **Clapper Lark**, and the stretch from here to Bredasdorp (especially around the main De Hoop turn-off) is excellent for **Stanley's Bustard**. If you wish to return to the N2, you can turn right at the fork 1 km later (see map p.58), and follow another gravel road to

Swellendam, along which there are also good numbers of **Agulhas Long-billed Lark** and **Blue Crane**. The latter is a fairly common sight throughout this region.

The whole of the Overberg region is good raptor country; regularly seen species include **Secretarybird**, **Martial Eagle**, **Lesser Kestrel** and **Black Harrier**. Common and characteristic species of the agricultural lands are **White Stork**, **Black Crow**, **Southern Thick-billed** and **Red-capped Larks**, **Capped Wheatear**, **Orange-throated Longclaw**, **Pied Starling**, **Pin-tailed Whydah**, **Yellow Canary** and, particularly in stubble fields, **Cloud Cisticola** (p.57*).

DE HOOP NATURE RESERVE

This reserve incorporates 36 000 hectares of lowland fynbos and coastal dunes east of Cape Agulhas, including a low, fynbos-clad mountain (Potberg) and a coastal lake. The cliffs on the southern side of Potberg mountain are renowned for hosting the Western Cape's last breeding colony of **Cape Vulture** (p.72*), while the coastal thickets of the lowlands offer access to such desirable endemics as the **Southern Tchagra** and **Knysna Woodpecker** (p.72*).

From Cape Town, De Hoop Nature Reserve is most easily reached by taking the

N2 national road as far as Caledon, where you turn right onto the R316 to Bredasdorp. In Bredasdorp, turn left (north) onto the R319, and turn right, 6 km later, onto the signposted gravel road. The turn-off to the reserve entrance (also signposted) is 30 km along this road (a particularly good one for seeing **Stanley's Bustard**), and the reserve gate a another 6 km further on. Allow at least three hours to reach the reserve from Cape Town. An alternative route is via Swellendam, along the birding loops described on p.64.

Access to Potberg mountain is not, as one might expect, through the reserve's main entrance. Instead, continue along the gravel road from Bredasdorp (see map, p.58), drive past the turn-off to the main entrance, and follow the signs for about 10 km to the parking area and environmental centre below Potberg's southern slopes. The parking area is an excellent vantage point from which to look for **Cape Vulture** soaring over the nearby slopes, and a half-hour's scan overhead is bound to turn up birds wandering from the colony just to the east.

The eucalyptus (blue-gum) plantation and mixed alien and indigenous thicket along the stream adjacent to the parking area hosts a surprising number of interesting species. **Knysna Woodpecker** occurs in the riverine strip of eucalyptus, but is as unobtrusive and difficult to locate as ever (p.72*). These alien trees have also hosted all three of the south-western Cape's honeyguide species (**Greater, Lesser** and **Sharp-billed Honeyguide**), along with a contingent of forest raptors, including **Black Sparrowhawk, African Goshawk** and **Gymnogene. Swee Waxbill** and **Southern Tchagra** lurk in the thicket between the parking area and the eucalyptus. **Cape Siskin** occasionally visit from the mountain slopes above. The pleasant Klipspringer Trail ascends to the summit of the Potberg (4 km each way), offering some good fynbos birding (**Cape Rockjumper, Orange-breasted Sunbird, Cape Siskin**) in addition to guaranteed views of overflying

Cape Vultures. The colony itself is not accessible to visitors, in order to protect the breeding birds from disturbance. **Hottentot Buttonquail** (see p.23) and **Striped Flufftail** occur on the southern slopes of the mountain, but, as ever, it is very difficult to see these ground dwellers.

De Hoop's main entrance gate is located on a range of limestone hills, from which the road winds down onto the lowland fynbos-swathed plains below. The fynbos is interspersed with open, pasture-like areas, relics of efforts at agriculture prior to the proclamation of the reserve. Chacma Baboon (*Papio ursinus*), Bontebok (*Damaliscus dorcas dorcas*), Eland (*Taurotragus oryx*), Cape Mountain Zebra (*Equus zebra zebra*), Angulate Tortoise (*Chersina angulata*) and **Ostrich** favour these pastures and are bound to be seen in good numbers in the vicinity of the turn-off to the reserve headquarters and rest camp (on the right, 4 km from the entrance). **Capped Wheatear** and, peculiarly, **Namaqua Sandgrouse** also inhabit these short-grass areas. The latter are likely to be heard calling during mid-morning, as they fly over on their way to water. In the open fynbos, look out for striding **Secretarybird** and **Southern Black Korhaan**, and quartering **Black Harrier** (p.57*).

De Hoop Vlei

Sand and sea at Koppie Alleen.

De Hoop is one of the few places in the Western Cape where **Horus Swift** is regularly seen, most often in the vicinity of the vlei or flying low over the vegetated coastal dunes – for instance, those around Koppie Alleen. The parking area at Koppie

The reserve office and rest camp are set among dense, gnarled milkwood thickets adjacent to De Hoop Vlei. This large, irregularly shaped lake can at times attract a huge number and an excellent diversity of waterfowl and waders, although this varies greatly between years and seasons. **Great Crested Grebe** is regular here, and occasionally breeds in large numbers. **Southern Tchagra** is shy but reasonably easy to find in the thickets around the camping area – its call is loud and conspicuous (De Hoop is one of the more westerly sites where it may be found, although it occurs closest of all to Cape Town in *Acacia karroo* thicket in the Karoo National Botanical Garden at Worcester). **Knysna Woodpecker** is surprisingly frequent around the vlei, but is characteristically tricky to find (p.72*): look in the thickets partitioning the camping sites. Common residents of the vlei-side thicket are **Bar-throated Apalis**, **Sombre** and **Cape Bulbuls**, **Southern Boubou** and, rarely, **Black Cuckooshrike**. The area around the vlei and reserve buildings is also one of the best places in the southwestern Cape to see **Pearl-breasted Swallow**, which often feeds alongside other hirundines such as **White-throated Swallow** and **Brown-throated Martin**.

WATCHING THE WHALES

A bonus for springtime birders at the De Mond and De Hoop nature reserves (see opposite) is the remarkable number of whales visible close inshore. The south-western coast of Africa is the favoured calving ground of Southern Right Whales (*Eubalaena australis*) seeking more temperate waters during the harsh Antarctic winter. During the winter and springtime months (July–November, but especially August–September), they can be observed close inshore all along the Western Cape coast. The MTN Whale Hotline (toll-free 0800 22 82 22) will keep you up to date with the latest sightings. Good sites for whale-watching are along the Atlantic coast from Postberg (p.50), the western side of False Bay (Boyes' Drive, above Muizenberg, is an excellent vantage point) and, most famously, Hermanus on the south coast. Indeed, the marine reserve of Walker Bay, which Hermanus overlooks, has been recognized by the World Wildlife Fund (WWF) as one of the world's top twelve whale-watching sites.

Southern Tchagra occurs in thickets.

Alleen (follow the signs on the main road from the entrance gate) is at the eastern edge of a vast sand-sea of pure white coastal dunes – a splendid place in which to wander aimlessly and absorb the landscape. The rockier coastline to the east of Koppie Alleen is frequented by good numbers of **African Black Oystercatcher** (p. 32*), together with more widespread coastal birds such as **White-fronted Plover** and a selection of migrant waders including **Sanderling, Turnstone** and **Grey Plover**. From July to November, many calving Southern Right Whales (*Eubalaena australis*) make their appearance along this stretch of coast and are easily seen.

DE MOND NATURE RESERVE

This reserve is greatly underrated as a birding site. Quite apart from offering some excellent birding (notably **Damara Tern, Southern Tchagra** and a splendid diversity of waders), the reserve is a beautiful spot, centred on the broad and placid estuary of the Heuningnes River and flanked by battlements of white dunes. To reach De Mond, take the R316 southwards from Bredasdorp and, after 10 km, turn right onto the 16-km long signposted gravel road to the reserve entrance. Park at the reserve gate, and take a look around the adjacent milkwood thicket for **Southern Tchagra** as well as more widespread coastal-thicket birds such as **Fiscal Flycatcher** and **Acacia Pied Barbet**.

Take the footpath that leads from the reserve buildings and across a suspension bridge over the river, before following the western bank of the estuary to its outlet into the Indian Ocean (this is the start of the scenic, 7-km Sterna Trail, which loops from the estuary mouth westwards along the beach before returning over the dunes to the reserve office). **Pied Kingfisher** hunt over the river while, in summer, **Common Sandpiper** potter along its banks and small numbers of migrant waders feed at the estuary edges and roost on the protruding islands just downstream of the bridge. In addition to such common species as **Curlew Sandpiper, Ringed Plover** and **Grey Plover**, one can pick out scarcer and more localized birds such as **Bar-tailed Godwit, Curlew** and, occasionally, **Terek Sandpiper** and **Mongolian** and **Greater Sand Plovers**. **African Black Oystercatcher** and **Caspian Tern** feed at the estuary mouth (the reserve protects an important breeding colony of the latter). There is usually a tern roost on the sandbanks: **Swift, Common** (summer) and **Sandwich Terns** are most common, but small numbers of the diminutive **Damara Tern** occasionally roost here or feed over the estuary mouth, primarily from November to March (see box, overleaf).

The estuary at De Mond Nature Reserve.

SWELLENDAM

Swellendam, straddling the N2 national road a two-and-a-half hour drive east of Cape Town, marks the westernmost occurrence of several bird species. It is also the third oldest town in South Africa, and consequently boasts numerous fine Cape Dutch buildings. Rather improbably, it was the capital city of a mini-republic for three heady months in 1795, when its citizens declared independence and appointed a president! The municipal campsite in Swellendam, in addition to offering agreeable camping and affordable cottage accommodation, is a good site for several scarce species. The small path that leads through the dense riparian vegetation lining the flanking stream may yield **Tambourine Dove**, **Brown-hooded Kingfisher** and **Olive Woodpecker**. Flocks of **Swee Waxbill** scatter autumn-leaf-like at the campsite edges. **Grey-headed Sparrow**, at the limit of its range, can be found in the trees around the cottages, where a pair of **Wood Owls** call nightly.

THE DAMARA TERN COLONY

De Mond is the site of one of only two Western Cape breeding colonies of this declining species (currently, only a handful of pairs breed). The colony is 9 km west of the estuary and is best reached by road. From the reserve gate, retrace your route for 11 km, turn left at the crossroads and continue until you join the tarred R319 to Struisbaai. About 5 km before Struisbaai, take an inconspicuous turn-off to the left, signposted 'Struisbaai Plaat'. A gravel track runs through dune thicket (where Southern Tchagra occurs) and forks twice. At the first fork, turn right; at the second, turn left. Park in the small parking area at the beach's edge, and walk to your left (northwards). The colony is usually within 100 m of the parking spot, on the flat, shell-covered dune slacks where the dunes meet the beach. Please keep your distance, so as not to disturb the breeding birds.

BONTEBOK NATIONAL PARK

On the plains to the southeast of Swellendam, along the Breede River, lies the Bontebok National Park. The signposted turn-off is on the N2 just east of the town, and the park entrance is a further 3 km along this untarred road. **Quail Finch**, a scarce bird in the Cape, occurs in moist depressions

Bontebok: a conservation success story.

between the N2 and the park gate, especially opposite the Swellengrebel airstrip, 3.4 km south of the N2. A small number of **Eastern Red-footed Kestrels**, a very scarce bird in the Cape, have regularly been observed in this vicinity in summer. Much of the park consists of low, fynbos-clad plains, enlivened by grazing Bontebok (*Damaliscus dorcas dorcas*: an antelope which once came precariously close to extinction but which is now flourishing), Grey Rhebok (*Pelea capreolus*) and Cape Mountain Zebra (*Equus zebra zebra*). Driving the few kilometres across the plains to the rest camp along the Breede River at the park's southern boundary, you might be disappointed by the apparent paucity of birds. It is well worth scanning the plains, however, for **Secretarybird** and **Southern Black Korhaan** (p. 57*). Look for the occasional distant white dot which is likely to be a displaying male **Stanley's Bustard** (p. 72*). An early start and a thorough search through the roadside scrub between the park entrance and the rest camp should produce **Clapper Lark** (p. 116*) and **Grey-wing Francolin**, the latter feeding nervously at the road edges. This is also one of the better areas in the Overberg for **Martial Eagle** and **Black Harrier** (p. 57*).

There is excellent birding close to the rest camp, which offers both camping facilities and caravans for hire. A short trail starts behind the information centre and winds westwards through acacia thicket. Another begins at the bottom of the campsite, on the riverbank, and leads to an aloe-clad hillside bedecked with sunbirds in the winter flowering season. **Olive Bush Shrike** may occasionally be seen along the former trail. Other notable species (in Cape terms) are **Klaas's Cuckoo**, **Cardinal Woodpecker**, **Lesser Honey-guide**, **Southern Tchagra**, **Grey-headed Sparrow** and **Streaky-headed Canary**. **Pearl-breasted Swallow** nest annually in the camp buildings and are easily seen during summer; other common birds of the campsite area are **Cape Bulbul**, **Bar-throated Apalis**, **Fiscal Flycatcher**, **Southern Boubou**, **Malachite Sunbird** and, feeding on the lawn edges, flocks of **Swee Waxbill**. You are bound to hear **Fiery-necked Nightjar** calling in the campsite at night, and are likely to see such river-loving species as **African Black Duck** and **Giant Kingfisher** along the Breede.

Immature Klaas's Cuckoo

GROOTVADERSBOSCH NATURE RESERVE

The extensive wilderness area of the Groot-vadersbosch Nature Reserve incorporates a 250-hectare indigenous forest, the largest in the southwestern Cape and certainly the region's richest in bird diversity. A number of more characteristically eastern species reach their western limit here, and most are not difficult to find with a little patience and persistence, particularly in spring when they are at their most vocal. There is also good birding along the disturbed forest edges and in the adjacent moist mountain fynbos. Sought-after endemics that can be found at Grootvadersbosch reserve include **Knysna Woodpecker** (p. 72*), **Knysna** (p. 32*) and **Victorin's** (p. 73*) **Warblers**, **Forest Canary** and **Cape Siskin** (p. 33*), as well a host of exciting forest specials such as **Narina Trogon** (p. 125*) and **Crowned Eagle**.

To reach Grootvadersbosch, take the N2 national road east of Swellendam for 11 km, then turn left onto the R324 (signposted 'Suurbraak/Barrydale'). Continue along this road, perhaps stopping to scan the lily- and reed-fringed pond on the left, that some-times hosts **Giant Kingfisher**, **African Rail** and **White-backed Duck**. **Pearl-breasted Swallow** is also very often seen along this road. Pass through the picturesque village of Suurbraak, checking any flowering coral trees (*Erythrina caffra*) for feeding sunbirds,

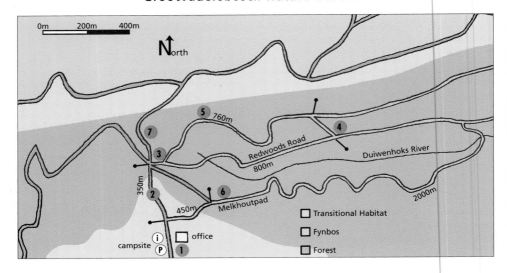

notably **Black Sunbird**. Where the road forks (the first fork is 26 km from the N2), follow the signposts to Grootvadersbosch (or 'Boosmansbos Wilderness Area'). The 'grootvader' (grandfather) of the forest's name was an 18th-century Dutch farmer to whom the land was first assigned. From Suurbraak to the reserve, the road passes over a series of rolling hills over which **Black Harrier** (p.57*) regularly hunt.

The reserve entrance (at ① on map, above) is on a ridge overlooking the forested valleys and the Langeberg range. Next to the entrance is a parking area, an information centre and a beautifully-situated campsite. Groups of **Cape Siskin** invariably forage around the campsite, and **Greater Double-collared Sunbird** feed in the garden of the office, where you need to obtain a day permit to enter the forest. From here it is a short, steep walk down a bracken-covered slope known as Bosbokrand ('Bushbuck Ridge'; ② on map) before you enter the forest on an excellent network of paths. West of the ridge, the forest is predominantly slightly drier and lower than that to the east of the ridge. A gentle walk of roughly three

hours (allowing time for lots of birding) is a 1.7 km loop through the eastern valley, described below. However, you might also want to make a short foray into the western portion of the forest, or into the moist, fynbos-clad slopes above.

Disturbed areas, such as those on the descent from the campsite to the forest edge at Bosbokrand, are the favoured feeding habitat of several seedeaters, including small flocks of **Swee Waxbill**, **Forest Canary** and **Cape Siskin**. This is also a good place to look for foraging **Black Saw-wing Swallows**, and for raptors. **Crowned Eagle** reaches its western limit at Grootvadersbosch, and is regularly seen overflying this ridge; more common forest raptors are **Forest Buzzard**, **Black Sparrowhawk** and **African Goshawk**. These are often seen perching on the skeletons of the introduced Giant Redwoods that protrude through the canopy at various places in the forest. **Cuckoo Hawk** has also been seen here in recent years.

From Bosbokrand, turn right at ③ onto the signposted 'Redwoods Road', a gravel track winding down the slope's lower contour. Walking down this track early in the morning, you are likely to catch glimpses of the Cape's westernmost **Red-necked Francolins**, scurrying off the path edge or calling tantalizingly fifty or so metres ahead. The most common and conspicuous birds in the forest are **Sombre Bulbul**, **Cape Batis** and **Bar-throated Apalis**. Before long, however, you will intercept a bird party and thus be likely to encounter **Olive Wood-pecker**, **Terrestrial Bulbul**, **Paradise** and **Blue-mantled Flycatchers**, **Yellow-throated Warbler** and **Greater Double-collared Sunbird**. While all of these are very vocal, some, such as **Terrestrial Bulbul** and **Olive Bush Shrike**, are inconspicuous lurkers, and you will need to invest a little time before obtaining good views.

After a gentle descent, the path turns sharply to the left and crosses the Duiwen-hoks River. The streamside undergrowth holds **Knysna Warbler**, typically vocal yet skulking as ever (p. 32*). This junction is also a good spot for **Knysna Woodpecker** (p. 72*), which is not uncommon in the forest, as evidenced by its very distinctive, shrieking call. The challenge to birders lies in that this species only calls at 10–15 minute intervals, though it can often be located by following its soft but more regular tapping. **Olive Woodpecker**, conversely, is both vocal and conspicuous.

Ignore the continuation of the broad gravel track at ④, and turn left along the foot-path back up the northern slope of the valley, where the forest is taller and moister. The forest clearing a few hun-dred metres up this path is a good spot to look for over-flying raptors. Also look out for **Grey Cuckooshrike**, a subtly beautiful canopy species that, though fairly common, requires a little alertness to its peculiarly sibilant call, vaguely reminiscent of that of **Dusky Flycatcher**. Another stunning canopy bird that reaches its western limit at Grootvadersbosch is the **Narina Trogon** (p. 125*). It is surprisingly easy to locate once you become familiar with its repetitive, hoarse hooting call. Listen for it in the vicinity of the canopy hide at ⑤, as well as elsewhere in the forest. Throughout the forest, look out for South Africa's most westerly Bushbuck (*Tragelaphus scriptus*), which bark startlingly as they make a crashing escape through the dense undergrowth ahead.

Returning to Bosbokrand, you might want to make a short detour to the other canopy hide at ⑥ (500 m down the sign-posted 'Melkhoutpad', that follows a higher contour down the same valley as the Redwoods Road). This hide provides an excellent vantage point from which to scan for raptors, which often perch on the skeletons of the introduced redwoods across the valley from this hide.

North of Bosbokrand, a gravel track (⑦ on map) soon leads into moist fynbos to become the beginning of the Boosmansbos Hiking Trail, a route that leads the rugged on a two-day loop among the Langeberg peaks. **Victorin's Warbler** (p. 73*) is very common here, even a few hundred metres from the forest edge. **Red-wing Francolin** also occurs on these slopes, but is decidedly scarce.

Grootvadersbosch, in a valley of the Langeberg Mountains.

SELECT SPECIALS

Cape Vulture

Depending on their itinerary, visiting birders touring South Africa would find it worth their while to attempt to see this species in the Overberg, despite only about one hundred birds remaining here. A common fallacy is that this bird is easy to see in the game reserves in the east of the country, but it is in fact difficult to find except in the Drakensberg mountains, in the far Eastern Cape Province, and in the vicinity of breeding colonies in the Northern Province. The Potberg colony (p.65) is notable as the last existing in the Fynbos Biome, and currently consists of just 32 breeding pairs. This species has declined drastically in South Africa in the past few decades, primarily due to poisoning from stock carcasses laid out by farmers to eradicate vermin.

Blue Crane

This is the country's national bird, although it escapes being a true South African endemic by virtue of a small population on Etosha Pan in northern Namibia. In the Western Cape, the birds occur fairly scarcely in the wheatfields of the West Coast's Swartland region (see p.45), but are common and easy to find in the Overberg wheatlands. They are especially conspicuous in winter, when breeding pairs congregate into large flocks of up to several hundred individuals. Indeed, this is one of the few rare birds that may have benefited from the destruction of lowland fynbos in favour of the agriculture that emulates its grassland habitat. Driving along the N2 national road east of Cape Town, you are bound to see cranes at the roadside – often close to small farm dams – even before reaching Caledon or Riviersonderend, and they may also be seen along the farmland loops described on p.63.

Stanley's Bustard (right)

Isolated in the moist grasslands and lowland fynbos of South Africa, Stanley's Bustard is currently classified as a subspecies of Denham's Bustard, whose range extends into East Africa. It has adjusted well to the Overberg wheatlands, particularly those east of Bredasdorp. Though never common, it is readily seen – refer to the main text for likely sites within the Bontebok National Park (p.69) and the Overberg farmlands (p.63). Worth looking out for during spring are the displaying males, which retract their heads, inflate their white throat pouches, and strut about in this voluminous state – an intriguing sight that, from a distance, one might uncharitably liken to a plastic shopping bag caught in the vegetation.

Knysna Woodpecker

This unobtrusive and little-known species is globally restricted to the narrow southern coastal strip of South Africa, extending as far north as the southern extremity of KwaZulu-Natal. It is not actually a very rare bird in its preferred habitat – dense coastal thicket or afromontane forest – but is rather challenging to see, as it calls only at lengthy and irregular intervals. The call is a short scream, likened by many to *'Skead!'* in honour of one of

the Eastern Cape Province's great natural historians, C.J. Skead. Closest to Cape Town, it can be found in the thickets of the De Hoop Nature Reserve (p.64) and the forests of Grootvadersbosch Nature Reserve (p.69). Further east, it occurs more commonly in the Garden Route forests (p.117).

Agulhas Long-billed Lark (below)

This species is a recent and localized split of Long-billed Lark (see p.13). It is restricted to the south coast, and is most common east of Bredasdorp (see Farmland Loops, p.63, for details of reliable sites). It is especially easy to locate in spring, when males regularly perform their looping display flights, launched from roadside fence posts and accompanied by a descending whistle. When not calling, feeding birds can usually be found in stubble fields, often near patches of roadside indigenous scrub.

Cape Rockjumper (above right)

Endemic to rocky areas in the fynbos-covered Cape mountains, primarily at higher altitudes, this species is most easily seen at Sir Lowry's Pass (p.60), Rooi Els (p.62), Bain's Kloof Pass (p.81) or, further east, at Swartberg Pass (p.123). However, it will be encountered by hikers on most of the Cape's mountain ranges. The presence of a group nearby is often first revealed by the piping alarm call of the sentinel bird. As you approach the outcrop where they are feeding, you will see the group members disappear between the rocks or move onto the next outcrop in low, gliding flight. Interestingly, the Cape Rockjumper breeds cooperatively: helpers (probably related birds) assist the parents in feeding the young, a behaviour that has attracted much research attention in this and other species.

Victorin's Warbler (below)

This fynbos endemic can be remarkably common in denser mountain fynbos. The call is faintly reminiscent of that of Grassbird but is more repetitive, and is, predictably, the key to seeing the bird. It can be maddeningly skulking, approaching to within a metre or two of playback but remaining well concealed in streamside thicket. The trick is to look for it in (or lure it into) slightly sparser vegetation and keep very alert to birds darting between denser patches and pausing momentarily before weaving into cover. Often, birds will pop up into clear view once, and then disappear to skulk obstinately thereafter, while tantalizingly continuing to call. Accessible sites are Sir Lowry's Pass (p.60), Bain's Kloof Pass (p.81), Harold Porter Botanical Garden (p.62), Grootvadersbosch Nature Reserve (p.71) and Swartberg Pass (p.123).

Tanqua Karoo Loop

'In the desert you have time to look everywhere, to theorise on the choreography of all things around you.' MICHAEL ONDAATJE, *THE ENGLISH PATIENT*

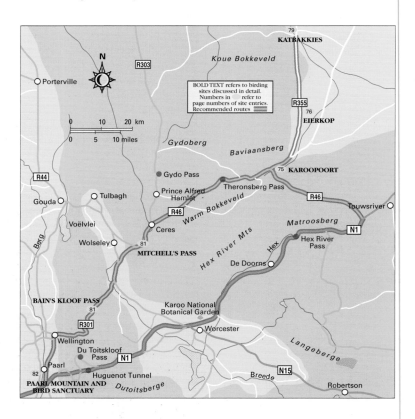

Visiting birders with little time at their disposal need not despair of seeing a good selection of the arid-country specialities of the Karoo semidesert. The majority are easily accessible within a day trip from Cape Town, and are set in some marvellous Karoo landscapes to boot. The parched brown expanses, aloe-lined escarpments and lonely isolated hills of the Tanqua Karoo provide an apt setting for such fine and sought-after dry western endemics as Karoo Eremomela, Cinnamon-breasted Warbler, Namaqua Warbler and Fairy Flycatcher, among many others.

TOP BIRDS

Karoo Korhaan, Karoo Lark, Southern Grey Tit, Tractrac Chat, Layard's Titbabbler, Karoo Eremomela, Cinnamon-breasted Warbler, Namaqua Warbler, Pririt Batis, Fairy Flycatcher, Black-headed and Protea Canaries.

Look for Karoo Eremomela on the plains below Eierkop hill.

The Tanqua Karoo, a part of the Succulent Karoo Biome (see p.7), merits one full day's exploration at the very least, although two days are preferable. As a dawn birding start is optimal, staying overnight in the nearby town of Ceres is a favourable option. This provides the freedom to explore, unhurriedly, both the Tanqua Karoo and the series of scenic and productive passes that lies between it and Cape Town. The route incorporating these passes is the less direct of the two main possibilities – the faster road, for a pre-dawn dash, takes in the N1 national road. It is quite feasible to make the round trip from Cape Town in a single day, but note, however, that this entails a total drive of about 500 km.

Visitors with limited time would do well to leave Cape Town about two hours before dawn and embark on the N1 to reach Karoopoort – at the edge of the Karoo – shortly after sunrise. From Karoopoort, you can work your way north to Katbakkies, stopping at the sites described below, before heading back again by mid-afternoon. Katbakkies makes a good lunch stop, as the birding here is not as dependent on early-morning activity as is that at Karoopoort and the plains between. You can then make a leisurely return to Cape Town via the scenic but undeniably slower 'three passes' fynbos route through the towns of Ceres, Wellington and Paarl.

The winelands town of Paarl has two excellent birding sites associated with it, offering, respectively, localized fynbos and waterbirds, and can be included in a Karoo excursion. Alternatively, these sites can easily be tackled as a relaxed day trip from Cape Town.

THE TANQUA KAROO: KAROOPOORT

The mere two and a half hours' drive from Cape Town to the Tanqua Karoo leads you through everything from the majestic peaks of the Du Toit's Kloof mountains (burrowed through by the 4-km Huguenot Tunnel) to the pleasingly geometric vineyard mosaic of the Hex River Valley. Fynbos grades into progressively drier scrub, and one ultimately emerges through a gap in the mountains onto the arid, scrubby plains of the Tanqua Karoo.

Before venturing into the Karoo, it is well worth stopping at Karoopoort, the gateway to the Karoo, for a number of dryland specials. To reach it, take the N1 from Cape Town and, 10 km before Touws River, turn left (north) onto the R46 (signposted 'Ceres/Hottentotskloof'). At the T-junction 33 km further on (75 on map, opposite), turn right (east) onto the R355. Along the R46, and particularly in the vicinity of this T-junction, tense coveys of **Grey-wing Francolin** are regularly seen feeding on the roadside in the early morning. From here, the R355 follows a reed- and thicket-lined riverbed, which passes though a gap in the mountains before reaching the open Karoo.

Fairy Flycatcher

Karoopoort to Eierkop

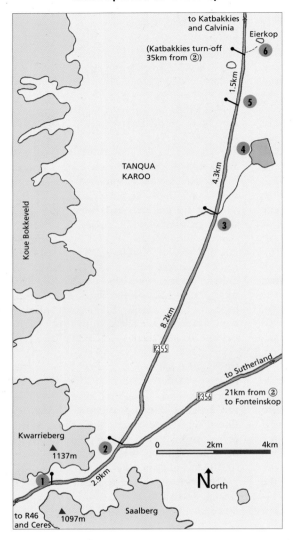

Among other typically dry western species of the acacia thicket are **White-backed Mousebird**, **Titbabbler**, **Fairy Flycatcher** and **White-throated Canary**. **Mountain Chat** occur on the rocky hillsides flanking the road and a pair resides along the first kilometre of gravel road. Flocks of **Red-winged** and, notably, **Pale-winged Starlings** fly purposefully overhead, balance on the cliff-faces, or feed on the fruiting fig trees next to the farmhouse. **Cinnamon-breasted Warbler** (p.85*) does occur here (especially at the picnic site at ②), but is more reliably found at Katbakkies, a little further north (p.79).

THE TANQUA KAROO: KAROOPOORT TO EIERKOP

Emerging from the Koue Bokkeveld mountains and onto the semidesert plains of the Tanqua Karoo, one soon reaches a fork in the road (② on map). In contrast with their unassuming appearance, each of these roads is associated with a South African record: the R356, to the right, leads to the town of Sutherland, which, thanks to its crisp, pollution-free desert skies, is the site of a world-class astronomical observatory. Rather less alluringly, it is ill-famed as the coldest place in the country! To the left is the R355 to Calvinia, similarly notorious as the longest road in South Africa uninterrupted by a town (250 km in all). The surface is of good

The essential Karoopoort species for many visitors will be **Namaqua Warbler**, which is a common and noisy bird of the *Phragmites* reedbeds and adjacent acacia thicket (see p.85*). A good area to look for it is opposite the oak-shaded farmhouse (① on map above). Though noisy, it sometimes requires a little effort to see.

Pale Chanting Goshawk

Looking south from the summit of Eierkop, with the bizarre succulent Tyledon paniculatus *in the foreground.*

quality gravel, but travellers driving all the way to Calvinia should nonetheless come well prepared with fuel and emergency water supplies – and should remember to check their spare wheel (see also p. 10). Birders should be particularly aware that braking suddenly on these roads could well result in a damaged tyre, as the gravel in this region is iniquitously sharp.

Common birds of the relatively moist scrublands just north of the road fork are **Pale Chanting Goshawk**, **Karoo Lark**, **Karoo Chat**, **Yellow Canary** and, more scarcely, **Sickle-winged Chat** and **Southern Black Korhaan** (p.57*). At ③, the road is crossed by an acacia-lined rivercourse where **Pririt Batis** (p.85*) and **Titbabbler** occur. You may notice the farm name 'Tierkloof' (see p. 8), referring to the Leopard (*Panthera pardus*) that still roam this region, although they are secretive and, sadly, now very scarce.

At ④ is a large dam, visible some distance to the east of the road. More often than not, the water level is low and the associated waterbirds are distant, reduced to amorphous shimmering blobs in the telescope. Nonetheless, it is worth the 200-m stroll down from the road to the farm fence to scan the water, as the adjacent scrub is, in any case, always good for a number of bird species. **South African Shelduck** and **Greater Flamingo** are often present on the dam, and **Namaqua Sandgrouse** occasionally fly in to drink. The dam is situated on the game reserve Inverdoorn; please respect the fence – there are rhino to enforce the law!

One kilometre further on, at ⑤, there is a prominent sandy intrusion on the landscape. A small group of **Anteating Chats** is usually present on these low, vegetated dunes. These most peculiar birds nest in burrows, hence their association with a soft substrate.

An excellent spot to look for several key Karoo specials is the distinctive pair of tillite hills straddling the road at ⑥. An inconspicuous gravel track, easily negotiable by two-wheel-drive, leads 500 m east from the R355 to the base of Eierkop, the right-hand hill.

Eierkop ('egg-hill') probably owes its name to the tiny, smoothed pieces of ostrich egg shells found on the summit; these were probably left by the early San hunter-gatherers, whose paintings are found in nearby rock shelters. Eierkop is arguably the most accessible site worldwide for **Karoo Eremomela** (p.85*). A small party is more often than not present on the plains surrounding this hill, moving quickly and inconspicuously from bush to bush, usually keeping an infuriatingly fixed distance ahead of their observers. Be sure to stay alert to their two calls (see p.85). Other common and typical birds of this habitat are **Karoo Lark**, **Karoo Chat**, **Rufous-eared Warbler** and **Grey-backed Cisticola**. The slightly denser scrub around the base of Eierkop supports **Southern Grey Tit**, **White-throated Canary**, **Malachite Sunbird** and, in spring during years of good rainfall, **Black-headed Canary** (p.105*) and **Lark-like Bunting.**

The short scramble up to the top of Eierkop is well worth the effort: the summit is bedecked with an intriguing diversity of succulents, and provides stunning panoramic views over the surrounding expanse of brown desert plains and shimmering purple mountains.

THE TANQUA KAROO: EIERKOP TO KATBAKKIES

As you head north from Eierkop along the R355, the landscape becomes progressively more arid until, approaching Katbakkies, bushes are few and far between and the

Tractrac Chat (above) is a plump, short-tailed bird with a pale rump, while Karoo Chat (right) is slender with a dark rump.

spot and legendary birding site popularly known as Katbakkies.

Another worthwhile detour on a return trip from Katbakkies takes in the less-travelled road linking the R355 to the R356. Five kilometres north of the turn-off to Katbakkies on the R355, turn right (east), onto an un-numbered road signposted 'Sutherland'. Before turning off here, it is well worth continuing north along the R355 for about 500 m and birding the patch of acacia thicket at a lay-by on the left-hand side of the road. **Pririt Batis** is almost guaranteed here, along with **Cape Penduline Tit**, **Yellow-bellied Eremomela**, **Titbabbler** and sometimes **Namaqua Warbler**.

ground gleams with the mineral patina of the desert pebbles. This is classic **Tractrac Chat** country: birds are most often spotted, 10–15 km north of Eierkop, as they flush near the road, and display their white rumps as they fly a short distance to perch again on a fence or low bush.

The commonest larks of this stretch of road are **Thick-billed** and **Red-capped**. **Spike-heeled Lark** is also regularly seen. It is worth keeping an eye out for pairs of superbly camouflaged **Karoo Korhaans**, although they have become scarcer here in recent years. Listen for their frog-like calls at dawn, and check in the shade of the occasional roadside tree at midday. Drainage lines with slightly denser scrub are good areas to search for small, restless flocks of **Cape Penduline Tit** (see box, p.81), best detected by their soft, inconspicuous call.

Pale Chanting Goshawks are reasonably common throughout the Tanqua Karoo, and **Greater Kestrels** frequently wander into the area. If you are lucky enough to visit after recent rain, you will see that pools forming close to the road invariably attract **South African Shelduck**, drinking flocks of **Namaqua Sandgrouse** and irruptive seedeaters such as **Lark-like Bunting**.

Twenty-one kilometres from Eierkop, turn left at the road signposted 'Kagga Kamma; Op-die-berg', to the small picnic

IDENTIFYING DESERT CHATS

The chats of the Karoo appear, at first glance, dauntingly alike. However, **Tractrac Chat's** much-discussed resemblance to **Karoo Chat** is only superficial: Tractrac is a smaller, paler and more compact bird, shorter-tailed and with a characteristically pale rump. The similar, but more scarce, **Sickle-winged Chat** can also have a remarkably pale, square rump-patch, but also appears longer-tailed and generally darker than **Tractrac**.

The un-numbered Sutherland road to the R356 offers a chance at **Karoo Korhaan**, **Greater Kestrel**, **Double-banded Courser**, **Namaqua Sandgrouse** and **Tractrac Chat**. When you reach the R356, turn right (south) and you will ultimately join the R355 at ② on the map on p.76. Fourteen kilometres south of the junction between the un-numbered road and the R356, there is a hillock (Fonteinskop) a little way to the west of the road which is also good for **Karoo Eremomela**. (Note, however, that this loop is about twice the distance of the direct return to Karoopoort on the R355).

The seemingly desolate Karoo plains offer great birding.

THE TANQUA KAROO: KATBAKKIES

The steep, rocky slopes and dense acacia thicket of the small picnic site often called Katbakkies, where the Karoo meets the escarpment of the Koue Bokkeveld ('cold buck-country') mountains, offers fine and varied birding. It is best known as the classic site for **Cinnamon-breasted Warbler** (p. 85*), a peculiar and often evasive warbler of arid, rocky hillsides. In addition, the acacia thicket offers several dry western specials, including **Pririt Batis**, **Fairy Flycatcher** and **Layard's Titbabbler**.

This site, which we shall call Katbakkies in keeping with popular birding tradition, is correctly named 'Peerboomskloof' or 'Skitterykloof'; the true, now ignored Katbakkies lies 20 km to the west. The obscure term 'Katbakkies' refers to the 'dickey-seat' on the rear of old motor cars, and the steep pass is so-named because it had to be ascended backwards, using the more powerful reverse gear. Just 3.5 km west of the R355, the road enters an aloe-lined gap in the mountains and proceeds to wind steeply up into the moister, fynbos-like scrub of the Koue Bokkeveld. Just past the initial passage into the mountains, a small track (①) on map overleaf) leads off to the left and immediately into the picnic site. A few concrete tables, a tap and a rustic toilet are set alongside impenetrable sweet-thorn (*Acacia karroo*) thicket that bristles with fearsomely huge thorns. Acacia trees line the riverbed leading into the top of the picnic site, where there is also a tiny, reed-bordered dam. Below the picnic site, the valley broadens and there is a small seep (at ②) that feeds a dense reedbed. The rock-strewn valley sides and precipitous, red-cliffed escarpments presiding over the picnic area are covered with a remarkable density of *Aloe comosa*, an unusually tall aloe restricted to a tiny area in this region.

These slopes are also legendary for being most accessible site in the world to see **Cinnamon-breasted Warbler** (p. 85*). Familiarity with this species' call is absolutely essential, as it is otherwise almost impossible to locate. The pair on the slopes adjacent to the band of red cliffs at ③, a short walk up the riverbed from the picnic site, are arguably South Africa's most tape-pressured birds, yet they nest here annually and still regularly respond to provocation. Having said that, we do urge ardent visitors to be very sparing with playback at this site in order to keep disturbance to a minimum. The birds tend to call loudly in short bursts, usually from exposed positions on a rock or aloe, then remain silent for another ten or fifteen minutes – so don't despair. Even during each bout of calling, they are constantly on the move, delving around the bases of bushes and scurrying rodent-like between rock jumbles.

There are several pairs of birds within a few hundred metres of the picnic site. For the sure-footed, another good area to

Cinnamon-breasted Warbler

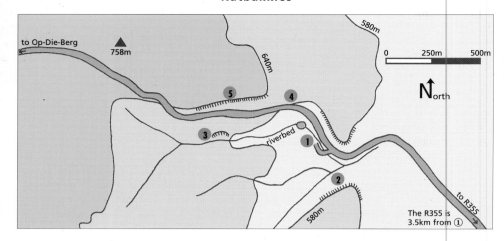

to Op-Die-Berg
758m
640m
580m
riverbed
0   250m   500m
North
The R355 is
3.5km from ①

Cinnamon-breasted Warblers may be found on the rocky slopes above the acacia-lined riverbed at Katbakkies.

investigate is the steep ridge that runs parallel to the right-hand side of the road. Begin at the bend in the road at ④, 150 m from the picnic site turn-off. Climb straight up the slope and on to the top of the ridge, aiming to skirt the right-hand edge of the cliff at ⑤. It is well worth the scramble, as this ridge has the advantage of keeping one above the birds, making them much easier to locate as they call intermittently among the rocks below. It is then possible to search for the birds rather than over-play the tape.

As you bound along the slopes after warblers, you're also likely to bump into **Southern Grey Tit**, **Layard's Titbabbler**, **Mountain Chat** and **Rock Martin**. **Pale-winged Starlings** regularly overfly the valley, and **Ground Woodpeckers** (p.105*) sometimes hurl invective from the ridges. **Dusky Sunbirds** occasionally move into the area. As ever, it is worth keeping an eye skyward for **Black** and **Booted Eagles** and **Rock Kestrel**.

The acacia thicket in the picnic site is usually alive with birds, even at midday. Essentials here are **Fairy Flycatcher** and **Pririt Batis**. Other interesting birds of this habitat are **Acacia Pied Barbet**, **White-backed Mousebird**, **African Marsh Warbler** (summer), **Yellow-bellied**

Eremomela, **White-throated Canary** and **Cape Bunting**. **Three-banded Plover** frequent the boggy lower seep, while the adjacent reedbeds host resident **Levaillant's Cisticola** and **Cape Reed Warbler**.

THE THREE PASSES: THERONSBERG, MITCHELL'S AND BAIN'S KLOOF

For those not keen to face the stresses of a dash back along the N1 to Cape Town, there is a very scenic alternative route via Ceres and Wellington. This takes in three superb mountain passes, all of which supply interesting birding in addition to marvellous mountain landscapes.

Theronsberg Pass, between Karoopoort and Ceres, has the gentlest landscape of the three, with grassy slopes frequented by **White-necked Raven** and, occasionally, **Black Harrier** (p.57*). Between this pass and Ceres are several small farm dams that are always worth a roadside scan for waterfowl.

Aloes are conspicuous on the rocky slopes at Katbakkies.

In Mitchell's Pass, west of Ceres, a good area to bird is the slope behind the conspicuous Tolhuis restaurant and pub. A footpath leads up from the shade of the Tolhuis oaks to a railway line on a protea-dense slope where **Protea Canary** (p.57*) may be found. **Bar-throated Apalis**, and occasionally **Swee Waxbill**, inhabit the dense bush just above the Tolhuis.

Bain's Kloof Pass, traversing the mountains above Wellington, takes one along 30 km of dramatic curves supported by dry-stone walls built, with the use of convict labour, some 150 years ago. The spectacularly rugged, boulder-strewn terrain is laced with icy streams, stained a deep tea colour by humic acid leached from herbivore-deterring plants. These streams support the dense vegetation favoured by **Victorin's Warbler** (p.73*), which is common in such habitat along the entire length of the pass.

At the summit of Bain's Kloof Pass there is a small settlement shaded by alien trees frequented by **Fiscal Flycatcher** and, more rarely, **Olive Woodpecker**. **Cape Rock Thrush**, often surprisingly scarce elsewhere, perch on the buildings. The ridges in this vicinity are good for **Cape Rockjumper** (p.73*) and **Cape Siskin** (p.33*). **Victorin's Warbler** also occur in the denser vegetation on the slopes. Scan overhead for **Black Eagle**. Birders caught out by nightfall would do well to carefully check all outcrops and telephone poles for the distinctive, bulky silhouette of the (admittedly ever-scarce) **Cape Eagle Owl** (p.105*).

NEST OF DECEPTION

The near-endemic **Cape Penduline Tit** is perhaps most famous for its nest, a deceptive structure equipped with a conspicuous predator-deterring false entrance (here seen directly opposite the belly of the bird). This leads to an empty chamber. To enter the concealed main compartment, the bird parts a hidden slit (with its foot, shown here opposite the bill of the bird) and slips in. The soft yet tough nest is made from fluffy wind-blown seed coverings or sheep's wool.

PAARL MOUNTAIN AND BIRD SANCTUARY

The oak- and jacaranda-lined streets of Paarl lie along the banks of the Berg River, at the eastern foot of the low, sprawling massif of Paarl Mountain. The massif offers good fynbos birding and is a well-known site for the inconspicuous and elusive **Protea Canary** (p.57*). A few kilometres north of the town centre is the Paarl Bird Sanctuary, a picturesque and productive sewage works that attracts an excellent

The Paarl Area

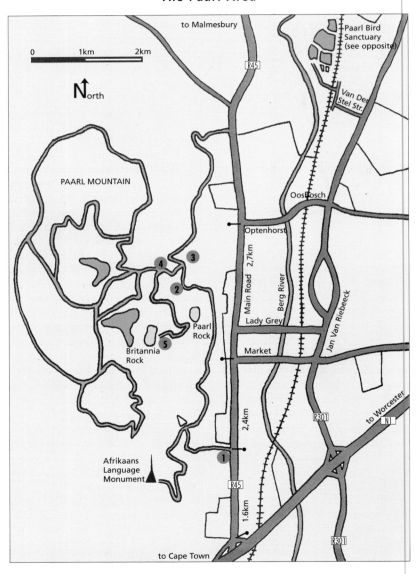

Paarl Bird Sanctuary

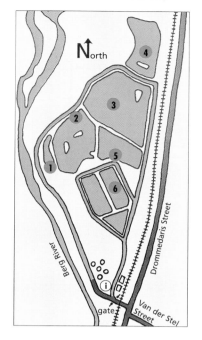

Granite domes and fynbos in the Paarl Mountain Nature Reserve. Malachite Kingfisher, below, is common in the Paarl Bird Sanctuary (overleaf).

diversity of waterbirds. These include several species that are otherwise fairly scarce close to Cape Town, such as **Little Bittern**, **Lesser Flamingo**, **White-backed** and **African Black Ducks**, **Water Dikkop** and **Malachite Kingfisher**.

Access to Paarl from the N1 national road from Cape Town can be confusing: there are two off-ramps to Paarl, and it is the first and rather inconspicuously marked 'R45: Main Road' turn-off that you need to take. The soaring monument to your left honours the Afrikaans language. The Paarl Mountain Nature Reserve, incorporating a small wild-flower garden in one of its wooded valleys, is easily reached from Main Road. At ①, turn left (1.6 km north of the off-ramp), onto Jan Phillips Mountain Drive.

The drive winds up through vineyards, then contours northwards, passing the wild-flower garden at ②, and the turn-off to the mountain reserve at ③.

The wildflower reserve is a small, pleasant botanical garden that often holds large numbers of confiding **Cape Sugarbirds** (p.33*) and **Orange-breasted Sunbirds** (p.33*). Other common species to be found here are **Black Saw-wing Swallow**, **Cape Bulbul**, **Bar-throated Apalis**, **Fiscal Fly-catcher**, **Cape Batis**, **Malachite Sunbird** and an assortment of canaries, including **Streaky-headed Canary** and, towards the top of the garden, the occasional **Protea Canary**.

Just beyond the wildflower reserve, at ③, you can turn left up into the Paarl Mountain Nature Reserve. At the entrance gate ④, a left turn takes you along a good gravel road to the base of Britannia Rock, whereas a right turn leads on to the network of roads that encircles the rolling plateau of the mountain top. The latter is recommended more for its scenic than its birding merits.

At ⑤, a path up to the 650-m crest of Britannia Rock has been cut into the granite, and, despite the often-present wind, it is well worth the climb. There are sweeping views of the Cape winelands and wheatlands set against the dramatic backdrop of the great ranges of Cape mountains that separate the coastal plain from the interior. Raptors such as **Black Eagle**, **Jackal Buzzard** and **Peregrine**

83

Falcon are often seen wheeling about the rock domes. Other notable birds of this area are **Ground Woodpecker** (p.105*), **African Black Swift** and **Cape Siskin** (p.33*). Although **Protea Canary** occur in denser stands of mountain fynbos throughout the massif, the bird is never easy to find.

The friendly, manageable character of the Paarl Bird Sanctuary (pictured below) contrasts strongly with the stark, windy nature of its Capetonian counterpart, the Strandfontein sewage works (see p.26). Paarl offers two well-positioned hides, a good diversity of habitats within a modest area, and almost guaranteed sightings of several tricky species.

Malachite Kingfisher, never an easy bird in Cape Town, is invariably present in secluded, reed-fringed inlets, such as those at the southern end of pan ①. This pan is also a good site for **White-backed Duck**, **Purple Gallinule**, and, in summer, **White-throated Swallow** and **Red Bishop** are much in evidence. It is also perhaps the best place in the Cape to see **Little Bittern**, which is occasionally observed perching at or flying along the reedbed edges. The excellent bird hide here was unfortunately burnt down, but there are plans to replace it.

The alien thicket adjacent to this pan is overrun with **African Marsh Warblers** and **Paradise Flycatchers** during summer. From here, drive on to the hide at the northern end of pan ②, which usually produces an excellent diversity of ducks. The hide holds special significance for South Africa's best known birder, Ian Sinclair, who ticked his landmark 900th species for southern Africa here, in the form of a vagrant **American Purple Gallinule**. In addition to the usual widespread waterbirds, several more interesting species can be seen here, among them **Southern Pochard**, **Maccoa Duck**, the occasional **White-faced Duck** (a bird that has been resident in the Cape for less than a decade, and is still scarce) and, oddly, the more typically mountain-stream loving **African Black Duck**, which visits from the adjacent Berg River.

Pan ③ is the closest place to Cape Town where **Lesser Flamingo** can be found, often flocking conveniently alongside the more widespread **Greater Flamingo**. In summer, keep a look out for the large flocks of fluttering **White-winged Terns**. In pan ④, an island supports a heronry where, among others, **Black-crowned Night Heron** breed. **African Fish Eagles** often roost in the gum trees at the far side, and **Common Sandpiper** and **Ethiopian Snipe** feed in the roadside ditch between pans ③ and ④. Returning towards the entrance gate, turn right to the hide at ⑤. The sandbank in front of this hide usually lures a good diversity of migrant waders, including the localized **Common Sandpiper**. The reeds here hold **Black Crake**, and **Baillon's Crake** has also been recorded. Before leaving, take a scan around the sewage mixing enclosures at ⑥, where a loose group of **Water Dikkop** usually eyes one disapprovingly.

SELECT SPECIALS

Karoo Eremomela

This is a fairly common but oddly inconspicuous Karoo inhabitant. A co-operative breeder, it occurs in small, agitated flocks that remain constantly on the move, thoroughly gleaning low bushes before the birds follow each other onwards. It calls often, and indeed this is the best means of locating this species. Note, however, that only one of the two principal calls can be heard on the available commercial recordings. The other common call is a rapid 'krrr-krrr-krrr', rather reminiscent of Spike-heeled Lark. The classic site for these birds is around Eierkop in the Tanqua Karoo (p.77), but they are in fact widespread throughout this region and Bushmanland. There are also good sites close to the towns of Brandvlei (p.90), and Springbok (p.98).

Namaqua Warbler

Formerly classified as a prinia, this species has recently been assigned its own genus, *Phragmacia*, picturesquely named after its habitat of mixed

Phragmites reeds and tall *Acacia* thicket. It is a much more secretive bird than the similar Karoo Prinia, but every bit as noisy. The closest place to Cape Town to see Namaqua Warbler is Karoopoort

(p.76). It can also be abundant along the Orange River reedbeds (such as those at Upington, p.111); in the Augrabies Falls National Park, (p.112), in the campsite at the Karoo National Park (p.123), at the Shell service station in Calvinia (p.89), and in thickets around Leeu-Gamka and Three Sisters on the N1 national road from Cape Town to Johannesburg.

Cinnamon-breasted Warbler

A reticent and little-known inhabitant of arid, rocky hill-slopes, the Cinnamon-breasted Warbler is peculiar enough to have been accorded its own genus. Its behaviour most closely resembles that of a shy and diminutive rockjumper, bounding about sun-baked boulders and calling fervently before inexplicably disappearing for long periods (see also p.79). Katbakkies in the Tanqua Karoo is undoubtedly the best-known site for this species, but it is also reasonably accessible over the whole of Namaqualand (p.97), in the Karoo National Park (p.123), the Augrabies Falls National Park (p.112) and the Akkerendam Nature Reserve (p.89).

Pririt Batis

The rather curious name 'pririt' becomes much clearer if one attempts to pronounce it with a haughty French accent! The species was in fact named onomatopoeically by the intrepid 18th-century ornithologist François le Vaillant. The call can indeed be likened to 'pree-ree', a low, descending whistle often repeated ten or more times. Pririt Batis is inconspicuous when not calling, but is otherwise cocky and inquisitive, and is readily seen working its way through thorn trees. It is the only batis species found over much of southern Africa's dry west, preferring acacia-lined riverbeds and arid woodland, and is reasonably common in such habitat throughout the Northern Cape Province.

Bushmanland

'Hark! Hark! The lark at heaven's gate sings'. WILLIAM SHAKESPEARE, *CYMBELINE*

Bushmanland is a vast and sparsely populated semidesert of stark beauty. Its stony plains are scattered with low bushes, punctuated by broken country and the occasional dune-field. The freedom of these open spaces will be a welcome respite for those wearied by the stresses of city life, and the dedicated birder will equally appreciate its wealth of highly desirable southern African endemics. It is most famous in birding circles for hosting one of the world's highest diversity of larks, with an amazing 14 species occurring regularly. Furthermore, Red Lark is a true endemic to Bushmanland, and Sclater's Lark and Black-eared Finchlark are most easily seen in this region.

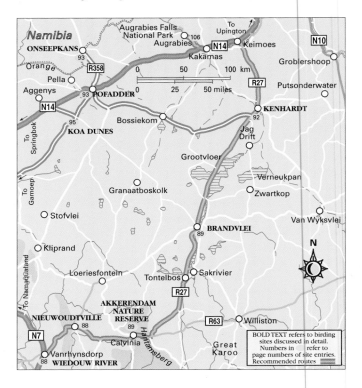

Bushmanland is a poorly defined area, bounded roughly by the Namaqualand highlands in the west, the Orange River in the north, and the towns of Kenhardt, Van Wyksvlei, Calvinia and Loeriesfontein in the east and south. Most of the rain, unlike that

Red Lark, a Bushmanland endemic.

TOP BIRDS

Ludwig's Bustard, Karoo Korhaan, Burchell's Courser, Red Lark, Sclater's Lark, Stark's Lark, Black-eared Finchlark, Karoo Eremomela, Cinnamon-breasted Warbler, Black-headed Canary.

in Namaqualand and the Cape Floral Kingdom, falls in summer. The best times of year for birding are the transitional seasons: bird activity is normally good in spring and autumn, and there is relief from the bitter nights of midwinter and the scorching heat of summer. Although many species here are nomadic and move around unpredictably in response to rain and seeding grasses, it is quite possible to find the majority during a short visit, using the sites described below.

Glancing at a map, you may be forgiven for thinking that this is a water-filled area, but all the 'blue lakes' shown are in fact relics of previous ages of plenty, and are today just shimmering, dusty expanses which hold water only in years of exceptional rainfall. Indeed, these promising blue patches are more suited for motor racing than any kind of aquatic activity – the parched and desolate expanse of Verneukpan near Kenhardt is rather improbably remembered as the site of Sir Malcolm Campbell's 1929 attempt, in his *Bluebird*, to break the world land speed record.

Besides the overwhelming feeling of freedom these seemingly endless open spaces provide, there are few tourist attractions in Bushmanland. Yet, despite the diversity of endemics found here and the fact that the region is highly regarded by discerning birders, it has been poorly treated by bird-finding guides and almost all the information given here has never been published before.

Bushmanland can be conveniently combined with a Namaqualand loop (see p.97), or even a visit to the Kalahari Gemsbok National Park (p.107). Those with limited time can see all the Bushmanland specials in the vicinity of the desolate little town of Brandvlei ('burning lake'), a mere seven hours' drive from Cape Town. However, those with an affinity for huge, near-empty landscapes and desert birds will enjoy three to four days here.

One excellent circular route is along the N7 north from Cape Town to Vanrhynsdorp, and then on east to Calvinia and up to Brandvlei. Bird around Brandvlei for a full day

A lone windmill stands sentinel above Bushmanland's expansive plains.

before heading north to Kenhardt, with the option of including a Kalahari Gemsbok National Park (p.107) loop at this point. From Kenhardt, travel west to Pofadder and Aggenys, where you can easily spend a day birding. Leaving Bushmanland, proceed into Namaqualand (p.97), spending a day around Springbok, with an excursion to Port Nolloth. Travel south through Namaqualand, perhaps incorporating some West Coast birding (p.41) en route to Cape Town.

Much of Bushmanland is partitioned off as private sheep farms. Good birding can be had at the roadside, but please ask permission before exploring farms. Away from the main arteries of the N14 and R27, there are long, desolate sections of unsurfaced road. Beware of travelling too fast on these deceptively safe-looking stretches, and please remember never to brake hard, even if there is a bustard at the roadside! Make sure that your spare wheel is in working order; an emergency tyre repair kit is recommended. Take plenty of extra water as the summer days can be exhaustingly dehydrating. Warm clothes are essential in winter, when night temperatures regularly drop below freezing.

CAPE TOWN TO VANRHYNSDORP

Heading north on the N7 from Cape Town, you will initially pass through the wheat-growing area of the Swartland (Malmesbury, Moorreesburg and Piketberg). The small patches of unassuming, greyish vegetation holding out between the wheatfields and on the lower hill slopes (best viewed from the Piekenierskloof Pass beyond Piketberg) are lowland 'renosterveld' (p.7), one of the most threatened vegetation types in the world. From a distance, the vegetation on the hill-slopes appears to be covered with large spots, which are in fact subterranean termite mounds (see p.103).

The Piekenierskloof Pass lifts you from the lowlands into the fynbos, and proteas line the road. Over the pass lies the fertile and intensively cultivated Olifants River valley, one of South Africa's main citrus-growing areas, and the fruit stalls along the road to Clanwilliam are well worth a stop. Dominating the eastern horizon are the majestic Cederberg mountains, the haunt of Leopard (*Panthera pardus*) and also of the rare Clanwilliam cedar tree (*Widdringtonia cedarbergensis*), endemic to this range. These mountains are also legendary in rock climbing and hiking circles. Ten kilometres before Clanwilliam, look out for the Paleisheuwel road (p.56), an excellent site for **Protea Canary** (p.57*).

Acacia karroo *trees line the Wiedouw River.*

Beyond the Olifants River, you leave the Cape Floral Kingdom behind and the fynbos gives way to progressively drier semidesert vegetation (see p.7). The petrol station at the entrance to the town of Klawer is an excellent place to refuel and take a refreshing break. Continuing towards Vanrhynsdorp, you may wish to bird the sweet-thorn (*Acacia karroo*) thickets at the Wiedouw River, 11.6 km north of the Klawer filling station. **Pririt Batis** (p.85*) is common here, and is best detected by its call. From Vanrhynsdorp, turn eastwards along the R27 (following the signs to Nieuwoudtville/Calvinia) and head towards the Vanrhyns Pass. Look for **Greater Kestrel** and **Black Crow** on the telephone poles in the open areas around Vanrhynsdorp (see p.104).

NIEUWOUDTVILLE TO CALVINIA

The dramatic Vanrhyns Pass abruptly lifts you above the arid plains of the Knersvlakte into moister vegetation again. Nieuwoudtville is world famous for its flowers, and the incredible density and diversity of bulbs produces a spring floral display that is arguably as spectacular as that of Namaqualand. Flowerwatching is at its best in the Nieuwoudtville Flower Reserve (3 km east of town on the Calvinia road) and at Glenlyon Wildflower Farm (ask for directions in town). Look out for **Clapper Lark** (p.116*), **Southern Black Korhaan** (p.57*) and **Grey-wing Francolin** in these areas. The Nieuwoudtville Falls, signposted 10 km along the Loeriesfontein road, are also definitely worth a visit. The scrub in the rocky areas surrounding the falls holds an interesting selection of birds, among them **Layard's Titbabbler**, **Fairy Flycatcher**, **Pale-winged Starling** and even the occasional **Protea Canary** (p.57*). **Cape Eagle Owl** (p.105*) occur here and in other rocky areas in the vicinity, but move around and are not easily seen. Continuing along the main tar road towards Calvinia and, ultimately, Brandvlei, you will notice

the landscape becoming markedly more arid, until you reach the semidesert of Bushmanland, north and east of Calvinia. Roadside birds to be on the alert for beyond Nieuwoudtville are **Ludwig's Bustard** (p.105*), **Booted Eagle**, **Black Harrier** (p.57*) and **Greater Kestrel**. At Calvinia, check the trees and reeds along the Oorlogskloof River, near the Shell service station, for **Namaqua Warbler** (p.85*) and **African Marsh Warbler** (summer). Turn into Hospitaal Street near the big silos and make your way to the nearby Akkerendam Nature Reserve. This is your last chance to see **Karoo Lark** (look on the plains shortly after you enter the reserve) as, in Bushmanland proper, only the similar but more localized **Red Lark** occurs. Park near the dam and walk along the broad path that leads into a huge amphitheatre. The hillside scrub along here holds **Layard's Titbabbler**, **Karoo Prinia**, **Karoo Robin**, **Mountain Chat**, **Cape Bunting** and **Black-headed Canary**, while **Fairy Flycatcher**, **Cape Penduline Tit** (p.81) and **White-backed Mousebird** prefer the taller vegetation along the river-course on the right. This reserve is also an excellent locality for **Cinnamon-breasted Warbler** (p.85*), which is best searched for on the rocky slopes to the left of the broad path. Scan the skies for **Black Eagle**, **Booted Eagle** and **Jackal Buzzard**.

BRANDVLEI: AN INTRODUCTION TO BUSHMANLAND

From Calvinia, take the R27 north to Brandvlei. As you enter Bushmanland, the landscape becomes more open and the bushes lower and sparser. From this point on, keep alert for **Black-eared Finchlark**, a nomadic species found throughout Bushmanland and which often moves around in flocks (p.96*). While driving, you are likely to spot the conspicuous, all-dark males fluttering over the road, although they invariably land frustratingly behind the

bushes by the time you have stopped the car!

About 54 km north of Calvinia (95 km south of Brandvlei; use the distance signs to orient yourself), check the pan on the left that often contains water and **Greater Flamingo**, **Black-necked Grebe**, **Avocet** and **South African Shelduck**. The

Rufous-eared Warbler

latter two may be seen on any pool of water in the region. **Blue Cranes** (p.72*) are regularly seen on the isolated patch of cultivated land 55 km south of Brandvlei, where **Thick-billed** and **Red-capped Larks** are common. There are colonies of **South African Cliff Swallow** (active September to April) under road culverts exactly 67 km, 97 km and 107 km south of Brandvlei.

Tractrac and **Karoo Chats** (see p.78) are the most common chats in the region, although **Familiar**, **Sickle-winged** and **Ant-eating Chats** may also be seen. The stocky **Chat Flycatcher** often perches on telephone wires, while **Karoo Long-billed Lark** (p.13), the region's most widespread lark, is often seen perched on the fence-posts. **Rufous-eared Warbler**, **Black-chested Prinia** and **Bokmakierie** are common on the scrubby plains, whereas **Cape Penduline Tit** (p.81) prefer drainage lines. Look out for **Ludwig's Bustard** (p.105*) and the occasional **Kori Bustard** (p.116*), which are often seen in flight, especially in the morning and early evening. The commonest raptors are **Pale Chanting Goshawk**, **Greater Kestrel** and **Lanner Falcon**, while **Martial Eagle** and **Black-breasted Snake-eagle** are also often seen.

Although **Red Lark** (p.96*) can be quite localized, it is found widely in the Brandvlei region and may be common where the vegetation is suitable. One such locality is precisely 23.6 km south of Brandvlei (1.1 km south of a picnic site), where a lone windmill

Taller shrubs: habitat of Red Lark.

desirable **Sclater's Lark** (see box) and **Burchell's Courser** (p.96*). From Brandvlei, take the gravel R357 towards Van Wyksvlei. Check the slightly thicker scrub (especially at ② and ⑤ on site map, below) for **Karoo Eremomela** (p.85*). Continue to the bridge at ③, where **European Bee-eater** breeds in summer, and where **Pririt Batis** (p.85*) occurs in the sparse riparian vegetation. Look for **Dusky Sunbird** around the occasional flowering bush (often on rocky koppies) and in riverine trees.

Shortly after the bridge, you will be faced with a staggered, three-point intersection at ④. Take the northerly road, signposted 'Van Wyksvlei'. The entire route from Brandvlei to Van Wyksvlei is excellent country for **Sclater's Lark**. In particular, check the habitat precisely at ⑦, and at 3 km beyond ⑦ along the R357. Much of the road towards Van Wyksvlei is also recommended for both this species and **Burchell's Courser**, and both can be found by driving along and stopping at

stands among a large tract of scrubby vegetation on the east side of the road. For the best chance of success, get here early and listen out for their calls. It is the 'plains form' of the **Red Lark** that occurs here, which is much browner than the richly-coloured 'dune form' occurring near Pofadder (p.95). **Karoo Lark** does not occur here. Also look out for the localized **Karoo Eremomela** (p.85*) here and in the scrub exactly 10 km south of Brandvlei. This is a relatively scarce species in Bushmanland, unlike the **Yellow-bellied Eremomela**, which is common over much of the region.

Brandvlei is a small town of exceptionally forlorn demeanour, situated on the plains of central Bushmanland. Its unprepossessing appearance is deceptive, however, as excellent birding may be had close to the town. Make an early start, leaving any non-birding companions to sleep late in the Brandvlei Hotel. **Red Lark** (p.96*) may be searched for at the site mentioned above, but it also occurs even closer to town: follow the R27 for 2 km north of town, and turn left along the unsurfaced road to Granaatboskolk. Continue for about 2 km along here, and check the scrub on either side of the road at ① (see map, right).

The area to the east of Brandvlei offers some excellent Bushmanland birding, including good sites for the

Brandvlei Area

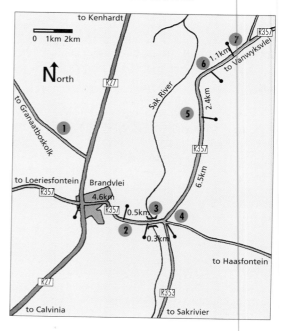

their characteristically open habitat. Check the small farm dams for **Sclater's Lark**. **Burchell's Courser** occurs sparsely here, and over the whole of Bushmanland. Watch for them running away from you in the distance when walking through suitable habitat, which is usually the most open, sparsely vegetated areas available – from open stony plains to the grassy edges of dry pan areas. Their unusual flight call often attracts attention. Look out for them while driving, too,

Double-banded Courser

about the pebbles and digging with their bills into the soft sand accumulated around the bases of bushes. **Lark-like Bunting** is often abundant. **Red Lark** occurs sparsely in the taller vegetation over this whole area: check at ⑥, and also at 17.7 km beyond ⑦ on the R357. **Namaqua Sandgrouse** are often seen flying to water in the mornings, delivering their characteristic, bubbling 'kelkiewyn' calls.

as their white wing patches are quite conspicuous in flight. Another rewarding route for both these species is the Haasfontein road: start from ④, and look especially on the plains around the farm Toekoms, 27 km from ④. **Red Lark** may be found 1 km east of ④.

You may also see **Karoo Korhaan** and **Double-banded Courser** at the roadsides, and **Black-eared Finchlark** is usually present in the area. A very common bird of the stony plains is the enchanting **Spike-heeled Lark**. These birds move about in small, active groups, jerkily scuttling

FINDING SCLATER'S LARK

This is a highly sought-after, nomadic species which is by no means guaranteed on a trip to the region (see also p.96*). Yet, despite varying in abundance locally, it remains fairly common throughout Bushmanland, and the dedicated birder does stand a good chance of finding the rather elusive bird. In this guide, we give as many reliable sites as possible, but the birds do move around, so the best strategy is to use these sites as starting points from which to embark on your own exploring.

As with many birds in this region, knowledge of their favoured habitat is essential: Sclater's Larks occur largely on very stony substrate with little vegetation. When walking through this habitat, be aware that their inconspicuous yet distinctive flight call is an excellent way to locate them. However, perhaps the best method of finding this lark is to wait patiently near a water trough or small dam adjacent to suitable habitat, and spend the otherwise unproductive heat of the day watching for birds coming in to drink, usually in pairs or small groups.

These open, stony plains are the habitat of Sclater's Lark.

BRANDVLEI TO KENHARDT TO POFADDER

The tarred route from Brandvlei to Kenhardt is good for most of the specials mentioned above, although birding is often more productive along gravel side roads. As one heads further north, the countryside becomes more grassy. After good rains, follow the side roads (for instance, 'Afdeling Pad 2984', on the right, 68 km north of Brandvlei) into the newly grassy plains, which often yield **Black-eared Finchlark** and, occasionally, in years of good rainfall, **Pink-billed Lark**. **Sclater's Lark** is also common in this area: wait for drinking birds at the farm dams 25–40 km south of Kenhardt. In particular, check the two dams

Water troughs are favoured by seed-eaters.

UNDISCOVERED CANARIES IN BUSHMANLAND?

While it is almost inconceivable that a bird species could to this day remain undiscovered in South Africa, an observation by a group of South Africa's most reputable and experienced ornithologists remains intriguingly unexplained. In late May 1989, while birding near the farm Jagdrift, south of Kenhardt (see map, p.86), Dr Peter Ryan and colleagues noticed a small party of four canaries. Both the males had a neat black face with a yellow supercilium, yellow underparts, green upperparts, and pale feather edges to the folded wing. Neither of the females had black faces; they had unmarked, grey plumage, except for their throats and bellies, which were white. No southern African or African canary fits this description, and the closest contenders are birds of forest edge further north in Africa. The encounter ended better for the canaries than for the ornithologists – by the time Peter had run to fetch his shotgun from the car, the birds had fluttered out of view and into birding legend.

(on either side of the road) 36 km south of Kenhardt, near the farm sign 'Knapsak'. Also look out for **Stark's Lark**, which is most often seen in the northern areas of Bushmanland. Bird the patch of trees that is exactly 100.1 km south of Kenhardt for **Titbabbler**, **Pririt Batis** (p.85*) and **Fairy Flycatcher**.

The town of Kenhardt is worth visiting for the kokerboom (*Aloe dichotoma*) forest, which lies just to its south. The giant nests of **Sociable Weaver** (p.110) may be seen on the telephone poles here; keep a look out for the associated and uncommon **Pygmy Falcon**, which also breed in the nests. From Kenhardt, one can head north to Upington and the Kalahari Gemsbok National Park (see p.107). To complete a Bushmanland loop, take the signposted road, 8 km south of Kenhardt, to Pofadder (marked on some maps as passing through Kraandraai or Bossiekom, although note that these are not towns). After 35 km along this road, check the stony desert plains near the intersection of powerlines for **Sclater's Lark**. This long and lonely road finally meets the R358, 27 km south of Pofadder.

POFADDER AREA

Pofadder, the stereotypical South African one-horse town, bears the name of the Puffadder (*Bitis ariens*), a sluggish and highly toxic snake found throughout South Africa. **Acacia Pied Barbet**, **Red-eyed Bulbul** and **Pale-winged Starling** occur in the town itself. The giant nests of **Sociable Weaver** are common in this part of Bushmanland. A particularly picturesque assemblage of these nests, with a resident pair of **Pygmy Falcon** (p.110), is found in some camelthorn trees exactly 11 km east of Pofadder (① on map below).

The 50-km unsurfaced road that leads from Pofadder to the Orange River at Onseepkans offers excellent birding across a diversity of habitats, from arid rocky gorges and dusty plains to lush riparian vegetation. The large-billed subspecies of **Sabota Lark** (see p.13), **Karoo Long-billed Lark** and **Karoo Chat** are common along the first section of this road. The acacia-lined

Below: Rosy-faced Lovebird

watercourse (②) crossing the road (6.3 km from Pofadder) contains **Acacia Pied Barbet** and **Pririt Batis**. After 15 km from Pofadder (from ③ onwards), check the open plains for **Burchell's Courser** (p.96*) and **Stark's Lark**, and

Pofadder to Aggenys

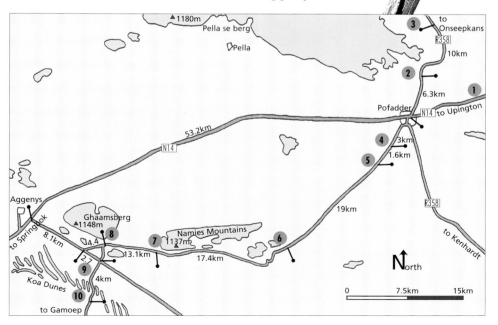

Karoo Korhaan

after 20 km look for **Scimitar-billed Wood-hoopoe**, **Sabota** and **Fawn-coloured Larks** in the low thorn-trees on the right-hand side. Thirty-three kilometres from Pofadder, a beautiful gorge falls away on the left of the road: look here for **Short-toed Rock Thrush** (scarce; see also p.124) **Cinnamon-breasted Warbler** (p.85*), **Pale-winged Starling**, **Dusky Sunbird** and **Black-headed Canary** (p.105*). As you reach the first buildings of the hamlet and border post of Onseepkans, check the group of palms on your right for breeding **Rosy-faced Lovebirds** (an extremely localized bird in South Africa) and **Palm Swifts**. **African Fish Eagle** and **Darter** frequent the river. The riparian vegetation offers **Swallow-tailed**

Bee-eater, **Cape Reed Warbler**, **African Marsh Warbler** (summer), **Namaqua Warbler** (p.85*), **Red-billed Quelea** and **Red Bishop**. The peachy-flanked subspecies of **Cape White-eye** (p.13) is common here, as is **Black-throated Canary**. **Fan-tailed Cisticola** display over the agricultural fields close to the Orange River.

From Pofadder, take the gravel road to the southwest, towards Namies. [This is the P2961 (not the R358), and can be reached by turning into Buitenkant Street (on the western edge of the town) from the N12. At the T-junction with Springbok Street, turn right and follow this road, which becomes gravel after a short while. Continue along the gravel and turn left into the P2961 after 0.3 km.] At ④ (see map, previous page), search the open plains, and check the water trough at ⑤ for **Sclater's Lark**. At dawn and dusk, listen for the frog-like duet of the **Karoo Korhaan**, one of the characteristic sounds of South Africa's arid plains. Continue through the Namies mountains, where **Cinnamon-breasted Warbler** occurs on the rocky slopes at ⑥ and ⑦. Be particularly careful when driving on the unsurfaced road in this vicinity, as there are a number of treacherously sharp corners. **Sabota Lark** (see p.13) and **Karoo Long-billed Lark** are common; look for the former in the taller scrub, where it draws attention to itself with its canary-like song. Also look out for overflying **Bradfield's Swifts**. At night, listen for the deep hoot of the **Cape Eagle Owl** (p.105*), which is resident on the slopes here, especially in the vicinity of the Ghaamsberg ⑧. This mountain is a treasure chest of biological diversity and threatened plants; sadly,

Grass covers the Bushmanland plains after good rain.

This rocky gorge near Onseepkans holds Cinnamon-breasted Warbler.

the proposal to build the world's largest open-cast zinc mine on its summit seems set to go ahead. Check the water troughs at ⑨ for large numbers of drinking **Namaqua Sandgrouse** and **Lark-like Bunting**. Continue to ⑩, where the road is crossed by intensely red dunes, and there is a stock enclosure lined with tyres on each side of the road. The dunefield is surrounded by sparsely-grassed and mesmerizingly flat expanses, punctuated by the occasional mountain. This is a well-known site for the richly-coloured dune form of **Red Lark**, which is in fact one of the most common birds in the vicinity. The larks are most easily found in the morning and late afternoon, when their rattling calls drifts across the dune crests. Between bouts of calling from the dune scrub, they feed on the ground, and can usually be seen running about between the bushes. Other common birds here include **Grey-backed Finchlark**, **Scaly-feathered Finch**, and **Anteating Chat**. **Fawn-coloured Lark** occurs less frequently. From ⑩, either retrace your steps to ⑨ and turn left towards the tarred N14 near the mining town of Aggenys (note that there is no accommodation available here) or, if well prepared with fuel and water, continue along the scenic gravel road to Springbok via Gamoep (see p.102).

The first 15 km or so of this road will prove rewarding for **Karoo** and **Northern Black Korhaans**; also check carefully for **Sclater's** and **Stark's Larks**.

If you have a spotlight, it is always worth taking a night drive along remote gravel roads: **Rufous-cheeked Nightjar** is common on the plains in summer, and **Cape Eagle Owl** may sometimes be seen near rocky areas. The Bushmanland also has an exciting diversity of nocturnal mammals (see box, p.124). Aardvark (*Orycteropus afer*), whose deep burrows can be seen throughout the region, is occasionally observed on night drives in Bushmanland. Other common mammals are Springhare (*Pedetes capensis*) and Bat-eared Fox (*Otocyon megalotis*); sadly, the latter is a regular road casualty.

Red Lark is common in the dunes at ⑩.

SELECT SPECIALS

Burchell's Courser

This endemic is undoubtedly the most frequently missed special in Bushmanland, and its elusiveness is attributable to its nomadic tendencies and low population density. The best places to search for it, and the recommended strategies, are given on p.90. In the non-breeding season, large nomadic groups may be seen in almost any open area. While today it is very much associated with arid areas from the Namib to the Karoo, the bird wanders widely, and small numbers move in winter into the higher-rainfall wheatlands of the south-western Cape, and into the grasslands of eastern South Africa. It was once even described as regular in western KwaZulu-Natal, although it no longer occurs there.

Red Lark

This little-known bird, whose nest was only discovered as recently as 1986, is the only species whose global range is totally restricted to Bushmanland. It is, however, reasonably common in its localized sandy habitat. Red Lark occurs in a number of colour forms, though recent research has shown that these are not sufficiently different from each other to warrant individual species status. Generally speaking, the rich reddish-backed 'dunes form' occurs on the red dunes in the northwest of the region near Aggenys (see p.95), while the browner-backed 'plains form' is found in the east of the region (around Brandvlei for instance; see p.89). The males engage in conspicuous aerial displays, during which they call frequently.

Sclater's Lark

Largely restricted to this region, Sclater's Lark is one of the enigmatic specials of Bushmanland. Suggestions on finding this inconspicuous bird, and its habitat preferences, are offered on p.91. It is unique among local larks in that it lays only one

egg, a phenomenon shared by several other un-related species (such as Double-banded Courser and Karoo Korhaan) that co-habit its harsh habitat. Parent birds undergo huge thermal stress while sitting on the nest, which is situated on exposed rocky plains. Once the egg has hatched, small stones are placed in the nest to break up the chick's shape and add to the camouflage. Although still locally somewhat nomadic, artificial stock-watering points on farms must have benefited this species as, conveniently for birders, it needs to drink regularly.

Black-eared Finchlark

This is a characteristically nomadic species that moves around in large numbers in response to rain and often irrupts into areas where the grass is seeding. Care needs to be taken in identifying the females, which may resemble other larks and finchlarks. Breeding must occur rapidly due to limited favourable conditions, and males hover butterfly-like during their character-istic display flights. The nest cup, lined with grass and distinctively surroun-ded by an earth and spider-web mix, is usually built at the base of a small shrub. Curiously for a lark, this species sometimes nests in loose colonies.

Namaqualand

'Full many a flower is born to blush unseen, and waste its sweetness on the desert air.'
THOMAS GRAY, *ELEGY WRITTEN IN A COUNTRY CHURCHYARD*

Namaqualand is best known for the spectacular spring floral displays that provide such a colourful yet ephemeral façade to a fascinating region. This winter-rainfall desert is home to a unique arid-land flora that is unparalleled globally in terms of its diverse mixture of both species and growth forms. The region forms the largest portion of the Succulent Karoo Biome (see p.7), recognized as the only desert biodiversity hotspot on earth and hosting the world's greatest variety of succulent plants. Yet, despite this floral uniqueness, Namaqualand shares most of its birds with the wider Karoo regions of Bushmanland and the Tanqua Karoo. Nonetheless, an excellent selection of endemics is available, and the region provides plenty of rewarding birding against a backdrop of floral richness and striking scenery. It is also the most accessible place in the world to see the endemic Barlow's Lark, and offers the best sites in the region for Ludwig's Bustard and Cape Eagle Owl.

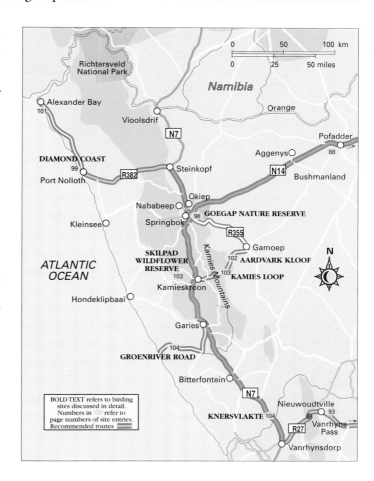

Namaqualand hosts spectacular floral displays in spring.

Namaqualand can be roughly divided into two very different regions. Inland lies a mountainous area, dominated by imposing granitic domes and jumbles of boulders. The N7 national road cuts northwards through these mountains, linking the string of small towns that serve the region. Between the uplands and the fog-shrouded Atlantic coastline lie the shrublands of the sandy lowland plains. The Orange River, bringing much-needed fresh water from the interior of South Africa, delineates Namaqualand's and South Africa's northern border with Namibia as it winds through the moonscape of the parched Richtersveld before spilling into the Atlantic Ocean.

Namaqualand is best visited during August and September, when the floral displays are at their peak (see box opposite). Overall, birding is best in spring and winter, and poorest in late summer. A visit is well combined with a loop through Bushmanland. Take the N7 national road from Cape Town, travelling north to Vanrhynsdorp (p.88). From here, you can follow a loop through Bushmanland (p.86), ultimately linking to Springbok from Pofadder, and continuing southwards through Namaqualand. Alternatively, you may choose to continue north from the Knersvlakte to Springbok.

GOEGAP NATURE RESERVE, SPRINGBOK

This 15 000-hectare reserve is situated in the rugged interior of Namaqualand, 15 km east of Springbok, the region's principal town. It offers spectacular flowers and scenery, and several Karoo specials, among them **Cinnamon-breasted Warbler** (p.85*) and **Karoo Eremomela** (p.85*), both found in classic Namaqualand landscapes. To reach the reserve (which only opens at 09h00), follow the signs from Springbok to the airport (via R355), and take another tarred road for a few kilometres past the airport to the reserve gates. There is an attractive succulent garden at the Goegap offices, which are also the starting point of a 17-km tourist loop road that covers a cross-section of the reserve's habitats (those suitably equipped should inquire about the exciting 4x4 routes, which allow one to explore more widely). Look in the vicinity of the offices for **Acacia Pied Barbet, White-backed Mousebird, Glossy Starling, Dusky Sunbird** and **White-throated Canary**.

Cinnamon-breasted Warbler is a common but inconspicuous inhabitant of rocky slopes throughout the reserve, and may even be seen on the hillsides behind the offices. Other birds characteristic of the rocky slopes

VIEWING THE FLOWERS

Dazzling floral displays are synonymous with Namaqualand: for a short period in spring, the region is carpeted with some of the world's most impressive shows. These usually occur from about the end of August to mid-September, but can be remarkably local and susceptible to rapid wilting, so contact Flowerline (see Useful Contacts, p.136) for the most up-to-date information. Rather ironically, these resplendent carpets are made up of annual species, which are in fact indicators of a disturbed environment. Discerning visitors might well take some time to wander away from the masses and examine the less abundant but just as impressive selection of flowering bulbs and succulents. Indeed, Namaqualand holds the world's greatest diversity of succulent species, spanning a mind-boggling array of growth forms, from the minute 'stone plants' (a quick look on the Knersvlakte, p.104, is recommended), to the giant kokerbooms (*Aloe dichotoma*) whose peculiarly geometrical silhouettes characterize the skylines of northern Namaqualand.

Cape Eagle Owl

of the reserve, which can be accessed at a number of points along the tourist loop (such as at 5 km and 8.5 km from the offices), include **Black Eagle**, **Booted Eagle**, **Ground Woodpecker** (p.105*), **Southern Grey Tit**, **Mountain Chat**, **Layard's Titbabbler** and **Dusky Sunbird**. **Cape Eagle Owl** (p.105*) is resident on the rocky slopes at Goegap, and is best searched for by driving along roads adjacent to the reserve at

night. In the more open areas along the loop road (at 4 km and at 8 km from the offices, for example), search for small groups of **Karoo Eremomela** among the low shrubs. Other birds which may be seen along here include **Ostrich**, **Ludwig's Bustard** (p.105*), **Karoo** and **Thick-billed Larks**.

DIAMOND COAST – PORT NOLLOTH TO THE ORANGE RIVER

To find the newly described **Barlow's Lark** (overleaf), you will have to descend from the baking and mountainous Namaqualand interior to the breezy scrublands of the coastal plain. Take the N7 from Springbok north to Steinkopf, checking for **Lanner Falcon**, **Jackal Buzzard** and **Black Crow** perched on the telephone poles. Note that the **Jackal Buzzards** in this area show a high incidence of variable white mottling on the breast, often causing confusion with the closely related **Augur Buzzard**, which only occurs much further north, in Namibia. From Steinkopf, take the R382 down Anenous Pass, which links the mountains to the coastal plain and eventually brings you to Port Nolloth. Because **Barlow's** and **Karoo Larks** are quite similar, it is worthwhile familiarizing yourself with the locally occurring subspecies of the latter. You will have this opportunity 5 km before reaching Port Nolloth, where **Karoo Lark** (reddish-brown upperparts; see box overleaf) occurs

Goegap Nature Reserve

Port Nolloth

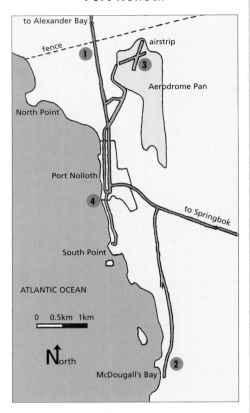

BARLOW'S LARK HYBRID ZONE

Barlow's Lark was only recently recognized as a full species (see p.12) after a study examined the group of closely related larks to which it belongs.

Close examination of these confiding birds reveals clear contrasts between Barlow's and Karoo Larks. Barlow's Lark is relatively easily distinguished by its clear, unstreaked flanks and heavier bill. Upperpart coloration is more complicated, as populations of both species on the white coastal sands show cold brown upperparts, which changes in the inland populations, on the redder sands, to a reddish-brown in Karoo Lark, and a sandy-peach colour in Barlow's Lark. There are also subtle differences in call. Despite the fact that Barlow's and Karoo

Lark are each well-differentiated species, a recent study discovered that they do hybridize over a narrow zone between Port Nolloth and Alexander Bay. The only place where this zone is publicly accessible is in the vicinity of Port Nolloth itself. The best way to distinguish hybrids is by flank streaking: pure Barlow's Larks show none of this (see picture above), while pure Karoo Larks show marked streaking. Hybrids typically show an intermediate streaking pattern, with faint scattered streaks on the flanks. Barlow's Larks tend to prefer the more arid, open habitat towards Alexander Bay and into Namibia, while Karoo Larks prefer the denser scrub found further inland, and their range extends southwards into much of Namaqualand.

commonly in the roadside scrub. Also keep a look out for Brants's Whistling Rat (*Parotomys brantsii*), which is quite common here as well as in sandy areas throughout Namaqualand. More frequently heard than seen, the rats draw attention to themselves by a soft, high-pitched whistle before disappearing down their extensive burrow systems.

Port Nolloth is a damp, salty town, often swathed in dense, rolling Atlantic fog. Most of the coastal strip of this region is frustratingly off-limits to birders (it supports rich alluvial diamond deposits).

Sparsely vegetated sandy plains near Alexander Bay are home to Barlow's Lark.
Opposite page, below: Southern Grey Tit

However, the most accessible site for **Barlow's Lark** is conveniently near the edge of town. Take the road towards Alexander Bay, and about 1.6 km beyond the buildings and the town's last tarred side-road, you will see on your left a fence demarcating the beginning of the mining area at ① (the fence is perpendicular to the road). Search the low coastal dunes on the Port Nolloth side of the fence for **Barlow's Lark** (cold brown upperparts; see box opposite), which is common here. However, note that this site is just on the edge of a zone of hybridization between **Barlow's** and **Karoo Larks**; be careful to distinguish pure birds from the much scarcer hybrids (see box). **Karoo Lark** (cold brown upperparts, see box) occurs again at McDougall's Bay, a few kilometres to the south of Port Nolloth (②). Other birds found in the scrubby strandveld vegetation in this area are **Cape Long-billed Lark** (p.13), **Southern Grey Tit**, **Cape Penduline Tit** p.81), and **Malachite** and **Lesser Double-collared Sunbirds**. **Bradfield's Swift** may be seen flying overhead anywhere in this area.

Port Nolloth can also be a good place to see **Damara Tern** (see p.68), which breeds in low numbers from November to February on the large pan on the northern edge of town ③. Seabirds and waders along the Port Nolloth coast at ④ include **Bank** (see p.21) and **Crowned Cormorants**, **African Black Oystercatcher** (p.32*), **White-fronted Plover** and, in summer, **Grey Plover**, **Turnstone** and **Sanderling**.

As you head through the diamond area towards Alexander Bay, rainfall decreases and the vegetation becomes lower and sparser. Check the telephone poles for **Black-breasted Snake Eagle**, **Jackal Buzzard**, **Pale Chanting Goshawk** and **Lanner Falcon**. **Tractrac Chat** occurs on the roadside. The private mining town of Alexander Bay has started to encourage ecotourism in the area and provides access to the mouth of the Orange, South Africa's largest river. Its estuary is an internationally-recognized RAMSAR wetland, offering good birding. The Northern Cape Nature Conservation Service is negotiating to include this rich area into a proposed transfrontier reserve. The more notable species here include **Greater Flamingo**, **South African Shelduck**, **Cape Teal**, **Maccoa Duck**, **African Fish Eagle**, **Avocet**, **African Marsh Harrier**, **Caspian Tern**, **Damara Tern** (uncommon), **Fan-tailed Cisticola**, **Cape White-eye** (see p.13), **Red-billed Quelea** and **Black-throated Canary**.

The surrounds of Alexander Bay are strikingly desolate, the monochrome landscape brightened only by the bizarre orange of the giant-lichen-cloaked hill adjacent to the turn-off to Alexander Bay. To find a pure population of **Barlow's Lark** (that is, away from the hybrid zone), continue past the turn-off to town, along the main road, which becomes unsurfaced as it swings inland along the Orange River. Check the area between Beauvallon and Brandkaros (10–20 km

Aardvark Kloof: the boulders hold Cinnamon-breasted Warbler, and Red Lark occurs on the sandy flats near the road. Below: Bokmakierie, a desert bushshrike.

beyond Alexander Bay), where **Barlow's Lark** (sandy-peach upperpart colorations, see box) is present in the sparsely-bushed areas on the right hand side of the road. Retrace your route to return to Springbok.

SPRINGBOK–KAMIESKROON: AARDVARK KLOOF

Aardvark Kloof is one of western South Africa's great endemic bird sites. Here, sandy spits from the open Bushmanland plains to the east meet the rocky Namaqualand interior, creating a mosaic of gentle, sandy-bottomed valleys flanked by boulder-covered slopes. At Aardvark Kloof, this diversity of habitats supports a bird community that will leave any desert-bird enthusiast twitching with indecision about where to look first. Here, the sharp calls of confiding **Cinnamon-breasted Warblers** (p.85*) echo through the roadside boulders, while **Red Larks** (p.96*) display a mere 50 m away!

Aardvark Kloof lies near Gamoep (see map, p.97), southeast of Springbok, and can be reached via the R355 (note that Springbok is your last source of petrol and

water). Follow the R355 straight past the final Airport/Goegap Nature Reserve turn-off, at the point where the tarred road turns to gravel. The unsurfaced roads in this region can be rather poor in places and should be negotiated with caution. Continue for 67 km beyond Springbok, to Gamoep, a small cluster of houses. Ignore the turn-off here (signposted 'Pofadder', 'Aggenys' and others) and continue for a further 2.6 km before turning right towards Kamieskroon. Follow this road for 2.1 km and bird the area just beyond the livestock grid in the road.

Red Lark is found in the small bushes on the right-hand side of the road. Such open areas (especially back towards Gamoep) support plains birds such as **Thick-billed** and **Karoo Long-billed Larks**, and **Karoo Eremomela** (p.85*). The rocky jumbles on the left-hand side of the road are home to **Cinnamon-breasted Warbler** and other endemics, including good numbers of **Southern Grey Tit**, **Mountain Chat**, **Layard's Titbabbler**, **Fairy Flycatcher**, **Pale-winged Starling**, **Dusky**

Sunbird and **White-throated** and **Black-headed Canaries** (p.105*). Small groups of **Ground Woodpecker** (p.105*) may be found on rocky slopes throughout the area. Scan the skies for **Black** and **Booted Eagles** and **Jackal Buzzard**. **Glossy Starling** is common here; **Pririt Batis** (p.85*) and **Acacia Pied Barbet** inhabit the acacia-lined watercourse on the right-hand side of the road.

Aardvark Kloof is also the start of an excellent scenic drive that winds its way back to the N7 at Kamieskroon, taking in spectacular landscapes that hold all the rock-loving hillside birds mentioned above (see map, p.97, although note that the road is unsurfaced and slow-going in places). You may wish to extend the scenic drive by turning towards Leliesfontein to head south through the Kamies Mountains and Studer's Pass before eventually arriving at Garies on the N7. Interestingly, these mountains, an elevated island of higher rainfall deep in the Namaqualand semidesert, support relict patches of fynbos (see p.5). Most of the land along the way is communally owned by the local pastoral people.

Although the mountains of central Namaqualand are a stronghold of **Cape Eagle Owl** (p.105*), we do not recommend extensive night driving along the unsurfaced mountain roads. However, you may wish to take a short nocturnal excursion along the gravel route that leads from Kamieskroon towards Leliesfontein (follow the signs from Kamieskroon). Scan for the owl on the telephone poles along the first 10 km of this road. Also listen out for **Freckled Nightjar**, which frequents rocky areas in this region.

The tar road between Springbok and Kamieskroon is by far the quickest way to travel through Namaqualand, and offers great scenery. Should you be in the position to drive this road at night, do check the roadside telephone poles and conspicuous boulders for **Cape Eagle Owl**, which occur in the rocky areas along the entire length of the road. **Karoo Lark** and **Karoo Chat** may be found in the flatter areas between the hills.

SKILPAD, KAMIESKROON

The Skilpad section of the newly proclaimed Namaqua National Park is one of the most popular flower-viewing localities in Namaqualand. Because most of the land is rather degraded (which enhances the displays of annual plants), birding in the reserve is poor, and it is not worth visiting out of season. However, the rocky slopes along the 21-km drive to Skilpad contain **Cinnamon-breasted Warbler** (p.85*), **Fairy Flycatcher** and **Layard's Titbabbler**. **Pririt Batis** (p.85*)

THE LITTLE HILLS

Travelling from Cape Town north through to Namaqualand, you will notice that, from a distance, the hill-slopes' natural vegetation often seems to be covered by a curious spotted pattern (see pp.88 and 104). Once a source of great mystery, the dots were later identified as the underground nests of the Harvester Termite (*Microhodotermes viator*). Measuring one metre high and 30 m across, many of these 'heuweltjies' or 'little hills' are thousands of years old. The reason that they appear as a pattern on the landscape is that, over the years, the activities of the termites have changed the nutrient composition of the soil, resulting in a change in the type of vegetation. This soil difference is most striking during spring, when the termite mounds are often clad with flowers of different colours to those of the surrounding areas (below).

occurs in the acacia-lined watercourses at the N7 road bridge near Kamieskroon and, in Skilpad itself, below the reserve offices. Keep an eye open for **Southern Black Korhaan** (p.57*), **Clapper Lark** (p.116*) and **Black-headed Canary** (p.105*) along the 5-km circular drive through the open plains of the reserve. To reach Skilpad, follow the signs from Kamieskroon, or enquire at the hotel.

GARIES TO THE GROEN RIVER MOUTH

Some of the most conspicuous examples of 'heuweltjies' (see p.104) are found just 4 km north of the town of Garies, on the hillsides to the east of the N7. The unsurfaced road that links Garies with the mouth of the Groen River (signposted 'Groenrivier-mond') is worth travelling in the winter months in search of **Ludwig's Bustard** (p.105*). Other birds to look out for include **Southern Black Korhaan** (p.57*), **Karoo Lark**, **Southern Grey Tit**, **Mountain Chat**, **Chat Flycatcher** and **Bokmakierie**. **Acacia Pied Barbet** and **Pririt Batis** (p.85*) occur in the scattered patches of trees. The Groen River estuary

---

HEAVISIDE'S DOLPHIN

This small, handsome cetacean (*Cephalorhynchus heavisidii*) is a distinctive endemic of the western coast of South Africa and Namibia and may be seen inshore along much of the seaboard (see pp. 46 and 55).

---

itself offers good waterbirds, including **Greater Flamingo**, **South African Shelduck** and **Cape Teal**. **African Black Oystercatcher** (p.32*) occurs along the coast, **Cape Long-billed Lark** (see p.13) is common in the coastal scrublands, and Heaviside's Dolphin (see box below) can often be seen just offshore.

THE KNERSVLAKTE

This lowland expanse at the southern edge of Namaqualand, which consists of a mosaic of quartz-strewn plains and sandy dunes, is perhaps best known for the diversity of miniature succulent plants which survive on the seemingly barren, rocky plains. It is definitely worth a brief roadside stop to see these bizarre succulents, large numbers of which can be observed adjacent to the N7 national road – for example, opposite the 'Douse-the-Glim' sign (22.6 km north of Vanrhynsdorp). The name Knersvlakte, or 'gnashing plains', probably refers to the crunching noise made by pioneers' wagon wheels as they crossed the expanses littered with quartz stones.

Greater Kestrel and **Black Crow** are regularly seen at the roadside in this area, their untidy stick nests conspicuously perched on telephone poles. Few birds are present on the stony plains, and the sandy areas offer the best birding. To reach an especially rewarding spot, travel along the N7 and, 14 km north of Vanrhynsdorp, take the road to the west signposted 'Soutfontein' (just before the Varsch River). Continue for 3.6 km. Among the more interesting birds you should see are **Namaqua Sandgrouse**, **European Bee-eater** (spring and summer), **Clapper Lark** (p.116*), **Karoo Lark**, **Spike-heeled Lark**, **Cape Penduline Tit** (see p.81), **Ant-eating Chat** and **Rufous-eared Warbler**. Check for **Pririt Batis** (p.85*) in the denser vegetation along the Varsch River. Note that from Vanrhynsdorp you can visit Nieuwoudtville (p.87) and, beyond that, Bushmanland (p.86).

SELECT SPECIALS

Ludwig's Bustard

This large, endemic bustard is restricted to arid areas, and undertakes often unpredictable seasonal movements. Ludwig's Bustards appear to follow rainfall productivity, and, in winter, move from the surrounding summer-rainfall areas into Namaqualand. The best regions to search for them are Namaqualand (winter and spring), Bushmanland and the Karoo. The birds are most often spotted in flight in the morning or evening. Widespread casualties are caused by flying birds colliding with electricity pylons. This has prompted research by Mark Anderson of the Northern Cape Nature Conservation Service to minimize these fatalities.

Cape Eagle Owl

This enigmatic owl is infrequently seen because it occurs at rather low densities in largely inaccessible mountainous terrain. It is restricted to very rocky areas, where it remains well concealed during the day and is seldom flushed because of its confiding nature. Its bark-like hooting calls during the winter breeding season are often the only indication of its presence, and only the fortunate will observe this species on a visit to the region. Its western strongholds are in the Namaqualand mountains (p.103), parts of Bushmanland (p.94), the Cederberg (p.56), and the interior of the Hottentots Holland Mountains down to Betty's Bay (p.63). It has been suggested that the South African populations may belong to a different species from the much larger 'Mackinder's' Cape Eagle Owl (*Bubo (capensis) mackinderi*), which occurs further north from Zimbabwe to Kenya.

Black-headed Canary

This highly nomadic and often gregarious canary, endemic to the western parts of southern Africa, can be surprisingly difficult to find. Although generally widespread in arid regions, it is most commonly located in rockier areas, especially if

there are seeding grasses in the vicinity. It is best picked up by its high-pitched flight call, or by waiting near suitable drinking points. Two colour forms occur, and although white-faced individuals have been classified as a separate species by some ('Damara' Black-headed Canary, *Serinus (alario) leucomaela*), this characteristic seems very variable and mixed flocks are often observed. Curiously, although it is has a strikingly different plumage, its calls closely resemble those of the Cape Canary.

Ground Woodpecker

Found in rocky areas throughout the region, this endemic is one of only three ground-dwelling woodpeckers in the world. It breeds in burrows, and feeds almost exclusively on ants. It is found in small groups, which are best located by their harsh, far-carrying calls. The best places to see Ground Woodpeckers are the Cape Peninsula (pp.19, 21), Namaqualand (p.99), Overberg (pp.61, 62), Tanqua Karoo (p.80), Swartberg Pass (p.123), Karoo National Park (p.124) and Kransvlei Poort (p.56).

Kalahari

'We looked down onto the deep heart of the desert and the empty bed of a broad, winding watercourse ... a remote world sealed with red sand, and spread out as still as the waters of a locked ocean.' SIR LAURENS VAN DER POST, *THE LOST WORLD OF THE KALAHARI*

The Kalahari Desert is a vast and almost unpopulated area, stretching from the northern region of South Africa's Northern Cape Province into central Botswana. It is not a true desert in the very strictest sense of the word, but the southern Kalahari's classic dune landscapes and broad riverbeds, lined with gnarled acacia trees, will appeal to anyone with even the remotest weakness for romantic landscapes. Inhabitants of the Kalahari Gemsbok National Park (the South African portion of the Kgalagadi Transfrontier Park, a vast conservation area straddling two countries) include big game such as Lion, Cheetah and Gemsbok, a notable diversity of raptors, and also a colourful selection of dry woodland birds.

To the south of the Kalahari runs the Orange River, a powerful passage of water cutting through the aridity of the Northern Cape en route from its catchment in Lesotho's alpine reaches to its destination on the desolate Atlantic shoreline (see p.101). Visitors approaching the

Kalahari by road from the south will need to pass through the regional centre of Upington, sprawling along the river's verdant banks. A little to the west is the Augrabies Falls National Park, where the

TOP BIRDS

Bateleur, Red-necked Falcon, Pygmy Falcon, Kori Bustard, Double-banded Courser, Burchell's Sandgrouse, Giant Eagle Owl, Bradfield's Swift, Monotonous Lark, Clapper Lark, Short-clawed Lark, Pink-billed Lark, Long-tailed Pipit.

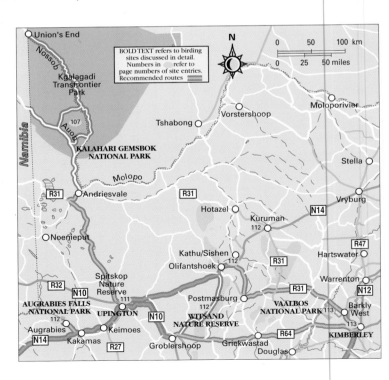

river plunges into a magnificent gorge that it has carved for itself through glistening granite. The birdlife here displays an interesting mix of Karoo and Kalahari elements, and well deserves exploration.

As one moves southeast through the vast and varied area between Upington and Kimberley, the provincial capital, the red sand, yellow grass and sculpted acacias continue. Here, a selection of bird species characteristic rather of South Africa's more wooded eastern regions add a tropical flavour to the birding. Our focus will not be on these more peripheral species, as they are much easier to find elsewhere, and we will concentrate rather on the specials of the region. In recent years, birders have been drawn to Kimberley in winter following the fascinating discovery of a new species of migrant pipit, which has added an exciting element of endemism to the already diverse birding.

Indeed, the Kalahari receives the majority of its tourists during winter, when the days are mild and cloaked by a resolutely blue sky, and the nights often bitterly chilly. In the Kalahari Gemsbok National Park, game is most visible at this time, and conveniently concentrated in the accessible riverbeds. During summer, the birdlife is augmented by a significant migrant cohort, and the resident birds are more active, especially after the late summer rains when the afternoon skies pile up with spectacularly dark and forbidding thunderheads that provide a dramatic backdrop to the lush veld.

Good roads (mostly tarred) link all the sites described below, and they can be easily combined with a loop through Bushmanland (p.86). A very bare minimum of two full days should be devoted to the park, although those with more time on their hands will find a week or more successfully spent. The Upington region and the Witsand Nature Reserve, in addition to being good birding spots in their own right, provide pleasant staging posts to break the otherwise gruelling full-day drives from Cape Town to the entrance of the Kalahari

Kalahari sunset framed by a camelthorn.

Gemsbok National Park, and from here to the Kimberley region. Visitors should note that Kalahari distances are vast, and should take particular care to allow sufficient travel time on unsurfaced roads (see p.10).

KALAHARI GEMSBOK NATIONAL PARK

This vast wilderness area offers the alluring combination of abundant game, superb landscapes and good birding, all within two-wheel drive access. The park is best known in birding terms for its remarkable diversity and abundance of raptors, including **Bateleur**, **Red-necked** and **Pygmy Falcons** and **Giant Eagle Owl** (p.116*). However, it also offers a number of other often underestimated specials, among them **Burchell's Sandgrouse** (p.116*) and **Pink-billed Lark**. Antelope such as Gemsbok (*Oryx gazella*) and Springbok (*Antidorcas marsupialis*) are common, and the park is also arguably one of the best places in the world to watch hunting big cats, like Lion (*Panthera leo*) and particularly Cheetah (*Acinonyx jubatus*). Spotted (*Crocuta crocuta*) and Brown Hyenas (*Hyaena brunnea*) also occur. The latter is actually the more common, but is rarely seen due to its crepuscular habits.

In 1997, the Kalahari Gemsbok National Park was officially joined to the Gemsbok National Park in adjacent Botswana to form South Africa's first Transfrontier Park, the Kgalagadi Transfrontier Park, a vast

The Kalahari is an excellent place to see Lion.

conservation area spreading over 3.6 million hectares of the southern Kalahari. It is now possible for visitors from South Africa to enter the Botswana portion of the park, provided that they are equipped with four-wheel-drive vehicles and have checked in at the Gemsbok/Bokspits border post, 60 km south of Twee Rivieren (see Useful Contacts, p.136).

The southern (South African) segment of the park forms a vast triangle enclosed by the Namibian border in the west, the Nossob River (also the Botswana border) in the north and east and, approximately, the Auob River in the south. These two rivers remain dry for decades on end, but are punctuated by numerous artificial waterholes that concentrate the game along the otherwise parched riverbeds. The principal roads in the park run along the length of the two riverbeds; the only other roads are two short-cut routes across the central dune sea

between the two rivers. The park roads are all unsurfaced, but are well maintained and fully accessible to sedan cars. The 260-km approach route from Upington is however tarred for all but the final 60 km of its length. There are three rest camps in the park: Twee Rivieren, at the entrance (the southeastern extremity); Mata-Mata, on the Namibian border in the far west; and Nossob in the north, halfway up the Nossob River.

The timing and length of your stay in the park will most likely depend on the availability of accommodation (camping facilities and very comfortable cottages are available in all the camps; see p.136 for reservation details), and on your own priorities. Because there are relatively few roads, and much of the park is uniform in character, the exact sequence is not especially important and the proposed route we describe below could be executed in any order. This route is an optimal birding one for those spending three to four nights in the park, and encompasses an itinerary from Twee Rivieren northwards to Nossob (day 1) and possibly Union's End (day 2), followed by an arc southwest to the Auob riverbed via the central dune sea, before returning to Twee Rivieren (day 2 or 3). Note that the drive between the Twee Rivieren and Nossob rest camps is deceptively long – allow at least half a day without stops. Enquire about current game-viewing

Camelthorn trees (Acacia erioloba) *line the dry Nossob River bed.*

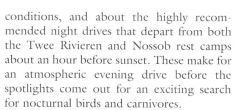
conditions, and about the highly recommended night drives that depart from both the Twee Rivieren and Nossob rest camps about an hour before sunset. These make for an atmospheric evening drive before the spotlights come out for an exciting search for nocturnal birds and carnivores.

Allow at least three and a half hours for the drive from Upington north to Twee Rivieren. If you have time to spare, the Spitzkop Nature Reserve just north of Upington is worth a brief visit (see p.111). Sixty kilometres from Upington, the landscape changes from open Karoo plains to sand, and the road takes a roller coaster route over parallel dunes. Watch for **Burchell's Courser** (p.96*) and **Black-eared Finchlark** (p.96*) along the initial Karoo expanses. **Northern Black Korhaan, Double-banded**

Courser and, in wet years, **Pink-billed Lark**, are all likely along the grassy dune sections of the road. Look out for the occasional **Pygmy Falcon**, which is dependent on the numerous **Sociable Weaver** nests ingeniously attached to the telephone poles (see p.110).

Near Andriesvale, 60 km south of the park, the road joins the confluence of the Molopo and Nossob rivers, which are wooded with giant camelthorn (*Acacia erioloba*) trees. This woodland, especially near Molopo Lodge, offers good roadside birding, including **Lilac-breasted Roller** and **Golden-tailed Woodpecker**.

KALAHARI GEMSBOK NATIONAL PARK: TWEE RIVIEREN AND BEYOND

At Twee Rivieren, have a look around the stand of thorn-trees to your right as you enter the main park gate for typical thornveld birds such as **Crimson-breasted Shrike**, **Scimitarbilled Woodhoopoe**, **Swallow-tailed Bee-eater**, **Kalahari Robin**, **Marico Flycatcher**, **White-browed Sparrow-weaver** and **Ashy Tit**. This is a good place to practise the productive birding technique of whistling the piercing call of **Pearl-spotted Owl** to attract mobbing bushbirds – you may well find yourself calling up an owl in the process, even during the middle of the day. In fact, all three of the Kalahari rest camps are excellent for owls, and a night walk to follow up on calls in any of the three is likely also to turn up **White-faced**, **Scops**, **Spotted Eagle** and **Barn Owls**. Ask the staff for their daytime whereabouts. **Pygmy Falcon** regularly hunts inside Twee Rivieren camp, and is in fact fairly common throughout the park.

We recommend spending the first night at Twee Rivieren, in order to make an early morning start northwards into the park. Shortly after leaving the rest camp, take the first turn-off to Mata Mata, and in the first 3 km of dunes look for **Pink-billed** and **Clapper Larks** (p.116*) and, if there have

The diminutive Pygmy Falcon.

attract **Violet-eared Waxbill** and its brood parasite, **Shaft-tailed Whydah**. **Red-necked Falcon** also starts to become much more common than further south in the park – look out for hunting birds targeting the flocks of drinking seedeaters; the falcons also regularly hunt over the Nossob rest camp.

If you are spending two nights at Nossob rest camp, it is possible to make the full day's excursion to the park's northern fingertip at Union's End. As you head north from Nossob and towards Union's End (look for **Burchell's Sandgrouse** at Cubitje Quap waterhole), bushveld elements such as **Lilac-breasted Roller**, **Yellow-billed Hornbill**, **Golden-tailed Woodpecker** and **Temminck's Courser** become more evident.

been good rains, **Kurrichane Buttonquail**. **Fawn-coloured Lark** is the park's most common lark and is also likely to be seen here, as is **Desert Cisticola**. Return now to the main Nossob road, and continue northwards past the junction of the Nossob and Auob riverbeds, aiming to reach the waterholes of Leeudril, Rooiputs and Kijkij 2–4 hours after sunrise. This will maximize your chances of seeing the scarce **Burchell's Sandgrouse** (p.116*), alongside the much more abundant **Namaqua Sandgrouse**. In wet years, listen for the incessant calls of **Monotonous Lark** along the river valleys.

All of the park's waterholes are invariably lined with flocks of drinking seedeaters making nervous forays to the water's edge, especially in winter. The most common of these are **Scaly-feathered** and **Red-headed Finches**, **Sociable Weaver** and **Lark-like Bunting**. During the summer months, huge flocks of **European Swift** are in evidence overhead, and **Lesser Grey Shrike** is commonly seen from the roadside. **Great Spotted Cuckoo** is also present at this time.

Continue towards Nossob rest camp. Dikbaardskolk picnic site, which makes a good lunch spot, is frequented by Ground Squirrels (*Geosciuris inauris*) and a selection of skinks and agamas. Look for the un-common **Great Sparrow** in the larger trees here and in Nossob rest camp. As you approach the camp, particularly northwards of Kaspersdraai, several additional species start becoming more evident: waterholes begin to

SOCIABLE WEAVER NESTS

The deceptively unassuming **Sociable Weaver**, a skilled builder, is responsible for the vast thatch structures that adorn many camelthorn trees and telephone poles in the Kalahari and northern Karoo. Each of these great avian apartment blocks is inhabited by up to several hundred birds, which use the nest primarily as a buffer against the extremes of desert temperature. **Pygmy Falcon** often appropriate nest chambers, and you can identify falcon-inhabited nests by the ring of white droppings around the chamber entrance.

In years of good rainfall, grassy areas along the dune roads support Pink-billed Lark.

The Union's End waterhole at the lonely junction of three national states – South Africa, Namibia and Botswana – is notable for occasionally hosting drinking flocks of **Rosy-faced Lovebird**. Nossob rest camp itself offers similar birding to Twee Rivieren. Ask the staff to show you the daytime roost of the resident **White-faced Owls**.

Two roads cut across the central dunefield between the two riverbeds. There is little game along these routes, but they are well worth the drive not only for the landscape and short-cut to the Auob, but also for **Pink-billed Lark**. This species is difficult to find in much of the Northern Cape and undergoes local movements. Look out for foraging birds around the bases of grass clumps, and do not be put off by their apparently white (not buff) outer tail feathers. Despite this incongruence with the field guides, they are indeed **Pink-billed** and not **Botha's Larks** (the latter has a localized distribution in eastern South Africa). **Northern Black Korhaan, Anteating Chat** and **Grey-backed Finchlark** are also particularly common along these roads.

Having reached the broad and shallow Auob riverbed near Kamqua waterhole, you have the option of either making a detour to Mata Mata rest camp on the Namibian border, or completing your loop through the park by returning to Twee Rivieren. The Auob riverbed is superb Cheetah country – enquire locally about game viewing. It is also an excellent place to see **Giant Eagle Owl** (p.116*); check the huge and gnarled camelthorn trees that line the riverbed.

UPINGTON AREA

Perched on the banks of the Orange River, Upington is the major centre of the north-western Cape. In this region, the river forms a verdant network, enclosing reed-fringed and now heavily cultivated islands that served as a 19th-century bandits' stronghold of some disrepute. Die Eiland Holiday Resort, situated on the south side of town, offers good birding, representative of this section of the river. To reach it from the town centre, follow the signs to 'Palm Avenue' (the longest palm-lined road in the southern hemisphere), which is the first turn to the left after you cross the river on the N14/N10 to Groblershoop. Birding in the riverine thicket along the river is likely to turn up Kalahari elements such as **Pearl-spotted Owl, Swallow-tailed Bee-eater, Ashy Tit, Red-eyed Bulbul** and **Brubru**, together with the more characteristically southerly **Namaqua Warbler**, and the strikingly dark race of **Olive Thrush** (see p.13). Keep a look out for **Abdim's Stork** in the vicinity (summer).

Riparian vegetation along the Orange River.

The Spitzkop Nature Reserve, 13 km north of Upington, consists of open grassy country and is excellent for larks. **Clapper** (p.116*), **Fawn-coloured** and **Spike-heeled Larks** are all common here; other birds include **Northern Black Korhaan, Pygmy Falcon, Chat Flycatcher** and **Anteating Chat**.

AUGRABIES FALLS NATIONAL PARK

The park can be reached along a 39-km road that leads northwest from the riverside town of Kakamas, which is 88 km west of Upington on the N8 national road. The focus of the park is the falls, and most visitors venture little further than the river. However, the surrounding plains and rocky outcrops provide good birding, offering such typical Karoo species as **Double-banded Courser, Ludwig's Bustard** (p.105*), **Burchell's Courser** (p.96*), **Spike-heeled** and **Karoo Long-billed Larks** and **Chat Flycatcher**. **Stark's Lark** and **Black-eared Finchlark** (p.96*) are erratic, and not always present in dry years. Two sought-after species of the rocky outcrops adjacent to the river are **Short-toed Rock Thrush** (look on the outcrops around the campsite) and **Cinnamon-breasted Warbler** (look around the viewsites along the gorge, such as those at Oranjekom, Swartrand and Echo Corner; p.85*).

The acacia thicket of the campsite at the park headquarters is also rewarding. Here, you can see **Golden-tailed Woodpecker, Acacia Pied Barbet, Ashy Tit, Red-eyed Bulbul, Namaqua Warbler** (common; p.85*), **Black-chested Prinia, Pririt Batis** (p.85*), **Pale-winged Starling, Dusky Sunbird**, and the attractively peachy-flanked Orange River race of **Cape White-eye** (see p.13). The short walk from the camp to the falls themselves may produce cliff-nesting species such as **Black Stork, Black Eagle** and **Peregrine Falcon**, as well as huge mixed flocks of aerially feeding swifts and swallows. You are likely also to see the multicoloured Cape Flat Lizards (*Platysaurus capensis*) that sun themselves conspicuously on the burnished granite.

WITSAND NATURE RESERVE

Witsand is a great birding site, named after its strikingly white reef of dunes that interrupt the red sea of the Kalahari sands. Adjacent to the dunes lies unexpectedly dense woodland and savanna, offering all the typical arid savanna birds of the Kalahari Gemsbok National Park (p.107), as well as species that prefer denser woodland. These include **Melba Finch, Black-cheeked** and **Violet-eared Waxbills, Yellow-billed Hornbill, Lappet-faced Vulture** and, in wet years, **Monotonous Lark**. Witsand is unique in hosting the only sandgrouse bird hide in the world: **Burchell's, Namaqua**, and the scarcer **Double-banded** may be seen drinking here; numbers vary, but are greatest in winter. Witsand offers pleasant camping and chalet accommodation, immaculately maintained by the Northern Cape Nature Conservation Service. See the map (p.106) for directions: it can be approached from either the north or the south (look out for conspicuous signposts just west of Olifantshoek, or 10 km east of Groblershoop). An isolated population of **African Rock Pipit** (p.125*) is found in the adjacent Langeberg mountains; turn east 13 km north of Witsand and follow the road to the Bergenaars Pass.

Northeast of Witsand, the small nature reserves adjacent to the towns of Kathu and Kuruman provide a host of woodland species and are worth visiting if you are passing through. Kathu is the most southerly place where **Red-billed Francolin** and **Pied Babbler** are regularly seen.

The Orange River gorge, Augrabies Falls park.

VAALBOS NATIONAL PARK

The park, 40 km northwest of Kimberley, incorporates a productive area of grassland and savanna, and is reached by a signposted left turn 19 km west of Barkley West. It also includes a stretch of Vaal River frontage, rich in alluvial diamond deposits. Indeed, mining and other pressures have reduced the park to the point where it may be facing deproclamation within the next few years. It remains well stocked with big game though, including Buffalo (*Syncerus cafer*) and Black (*Diceros bicornis*) and White (*Ceratotherium simum*) Rhinos.

Vaalbos's main attraction for birders lies in a largely isolated population of **Short-clawed Lark**, more typically a bird of overgrazed countryside in southeastern Botswana. At Vaalbos, it is uncommon: search for it where the open grassy plains are punctuated by small acacia trees, such as those near Block Dam (a map is obtainable at the gate). The open savanna areas around Block Dam are worth searching for **Buffy Pipit** and **Rufous-naped Lark**, while the nearby patch of trees holds such typical arid savanna species as **Brubru**, **Shaft-tailed Whydah**, **Crimson-breasted Shrike**, **Pririt Batis** (p.85*) and **Golden-breasted Bunting**. Wooded habitats throughout the park hold **Red-crested Korhaan** and, scarcely, **Little Banded Goshawk**. Grassland, such as that below the southern powerlines, offers **Northern Black**

Witsand Nature Reserve

Korhaan, **Anteating Chat**, and **Clapper** (p.116*) and **Spike-heeled Larks**. The Vaal River banks, accessed from the Riverside picnic site, offer **African Black Duck**, **Giant Kingfisher** and **White-fronted Bee-eater**.

KIMBERLEY AREA

Kimberley is often underestimated as a birding destination, despite entering the birding limelight in 1996 with the discovery here of a pipit new to science (**Long-tailed Pipit**, see box overleaf). It also offers access to several other species that are challenging to see in South Africa, such as **Bradfield's Swift**. A number of species that occur here, such as **Red-breasted Swallow**, **Crested Barbet** and **Golden-breasted Bunting**, are more characteristic of South Africa's eastern regions, and reach their western point of distribution in this area.

Kimberley is renowned as the site of a 19th-century diamond rush of unprecedented madness, one that converted a small hillock to what is now known as the Big Hole, a massive, water-filled pit gouged into the earth's surface. Today, despite its status as the industrial and administrative centre of the Northern Cape Province, the city is surrounded by natural areas and flanked on virtually all sides by private game farms that offer good dryland savanna birding.

The city proper offers access to two sought-after birds: **Bradfield's Swift** and **Long-tailed Pipit**. Look for the former at the Big Hole or De Beers Mine hole (follow the signs from the N12) where they breed, or observe them flying overhead anywhere in the city. **Long-tailed Pipit** is best sought at its type locality, Beaconsfield Park (also known as Keeley Park): follow the signs to the McGregor Museum and, leaving from the gate of the museum, go left and take the first left into Du Toits Pan Road and then the first right into Pratley Road. Continue to the big gates at the end of the road. Enter the park on foot, and turn left towards the series of playing fields, where

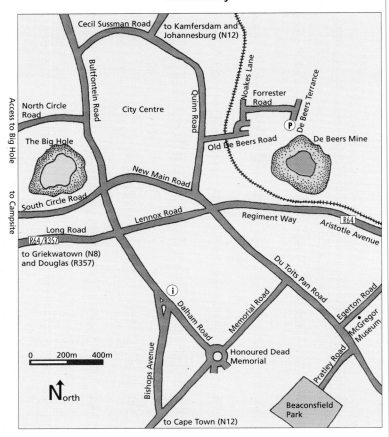

you will see pipits of all descriptions – including, potentially, **Long-tailed** (winter), **Plain-backed**, **Buffy**, **Grassveld** and **'Kimberley'** (see box). The thorn trees between the fields and the gate support **African Hoopoe**, **Lesser Honeyguide**, **Acacia Pied Barbet**, **Fiscal Flycatcher**, **Cape White-eye** (p.13) and **Black-throated Canary**. **White-backed Vultures** can at times be seen soaring over the city.

Access to most of this private land is limited. In addition to birding the roadsides, you may consider visiting Marrick Game Farm, which welcomes day visitors and also provides accommodation. It is conspicuously

signposted, 11 km west of Kimberley on the R357 to Douglas (tel: (053) 861-1530). Marrick offers a good combination of both savanna and open grassland, as well as an ephemeral vlei that occasionally hosts large numbers of waterbirds.

The woodland abounds with typical acacia thornveld species such as **Scimitar-billed Woodhoopoe**, **Ashy Tit**, **Kalahari Robin** and **Crimson-breasted Shrike**. Also present is the more easterly, small-billed subspecies of **Sabota Lark** (p.13). The open grasslands near the pan supply an entirely different selection of species, including **Northern Black Korhaan**, **Double-banded Courser**

De Beers Mine hole

KIMBERLEY'S NEW PIPITS

In the early 1990s, Dr Richard Liversidge, of the McGregor Museum in Kimberley, started noticing peculiar pipits around the town, and in 1996 the description of **Long-tailed Pipit** *Anthus longicaudatus* was published (see p.135).

The distribution and movements of this new species are poorly known, although it seems to be a non-breeding winter visitor (May to early September) to the Kimberley region. Pipit field identification is notoriously subjective, and separating **Long-tailed** from the similar **Plain-backed** and **Buffy Pipits** is less than clear. You will need to spend some time familiarizing yourself with the selection at Beaconsfield (see opposite) and elsewhere before attempting to sort them out. The publication of the description of a second new pipit – the '**Kimberley Pipit**' – is imminent, and visitors are cautioned to have a close look at all of the similar **Long-billed Pipits** around Kimberley. For updates, visit the African Bird Taxonomy page at www.birding-africa.com.

Identification: **Long-tailed** is a large, heavily built pipit with a distinct eyebrow and an unstreaked buffy back, crown and mantle (unlike **Grassveld**, **Long-billed** and '**Kimberley**'). Unlike **Buffy Pipit**, it has a distinctly yellow base to its lower mandible, like **Plain-backed Pipit**. Other subtle characters that may separate it from **Buffy** and **Plain-backed** are its slightly longer tail and darker colour, more horizontal jizz when feeding, and an even higher rate of tail-wagging, involving the entire lower body.

(common here), **Anteating Chat**, **Clapper**, **Spike-heeled** and **Red-capped Larks** and sometimes, with careful checking, **Pink-billed Lark**. Search the open parkland for the uncommon **Orange River Francolin**, where tall, scattered acacia trees stand above the grassy flats. This species, which calls in the early morning and is responsive to playback, displays a remarkable capacity to remain concealed. Although reasonably common in the Kimberley area, this bird is perhaps better looked for in the Free State or North West provinces (for example, in the very productive Sandveld Nature Reserve near Bloemhof), where it is more common and sites are more accessible.

Just north of Kimberley, on the N12 to Johannesburg, lies the vast expanse of Kamfer's Dam, one of the few perennial waterbodies in the Northern Cape and, as such, supporting an exceptional number and diversity of waterbirds. Access is possible through the golf course on the southern side of the pan, or from the adjacent N12 road. Waterbirds include **Greater** and **Lesser Flamingos**, **Black-necked Grebe** and small numbers of **Chestnut-banded Plover**. Reedbeds support **Golden Bishop**, here at the edge of its range. **South African Cliff Swallow** and, in wet years, **Black-winged Pratincole** feed over the adjacent grassland.

Kori Bustard

Famed as the heaviest flying bird in the world (although contested by certain Mute Swans), the Kori Bustard is remarkably common in the Kalahari Gemsbok National Park and is sure to be encountered in the riverbeds. Early settlers of Dutch origin, whose initial knowledge of natural history was largely based on tales of previous European exploration in Southeast Asia, often named African animals after these exotic beasts. Hence, leopards were called 'tygers' ('tigers', see p.8) and the Kori Bustard was named 'pou' ('peacock') because of its gaudy display, when the male inflates a feathery throat-pouch and splays its tail feathers. The Afrikaans name Gompou ('gum–peacock') refers to the bird's curious habitat of feeding on gum oozing from the bark of camelthorn trees. These majestic creatures also occur more scarcely in the Augrabies Falls National Park, and indeed throughout Bushmanland and the Kalahari.

Burchell's Sandgrouse

This near-endemic can be quite elusive, and is best seen in the Kalahari Gemsbok National Park (p.110). They are quite specific in their habitat requirements, and are only found on the red Kalahari sands. The best way to see them is to wait near a waterhole, especially between 2 and 4 hours after sunrise (p.110). Here a male is pictured soaking his absorbent, water-retaining belly feathers, which will provide moisture for his thirsty chicks upon his return to the nest.

Clapper Lark

This endemic is named for its rapid wing-clapping display (conspicuous in spring and summer) when the wings are beaten together a remarkable 26 times per second. Although it superficially resembles just another of Africa's brown larks, closer examination reveals exquisite, richly marked upperparts. The bird pictured here is the inland grassland subspecies that occurs in the Kalahari: the birds in the southwestern Cape show darker brown upperparts that contrast more with the orange underparts (see photograph on p.64). Taxonomic comments are provided on p.13.

Giant Eagle Owl

Despite its widespread presence in Africa, this species is rarely easy to find and will be a target of many birders visiting the Kalahari Gemsbok National Park. The best place to look for it is along the broad, dry bed of the Auob River, where it roosts by day in the huge, gnarled camelthorn trees. By driving along and scanning the trees for large silhouettes on the heavier branches, you are bound to pick up at least one or two on the drive from Twee Rivieren to Mata Mata. One dead giveaway is the presence, under trees, of small piles of neatly peeled-off spiny skins of the Southern African Hedgehog (*Atelerix frontalis*), which is a favoured prey item.

Garden Route and Interior

*'On the southern and southeastern coasts there are some fine forests ... the traveller
may also pass for days together through open plains.'*
CHARLES DARWIN, *A NATURALIST'S VOYAGE IN HMS BEAGLE*

As one travels eastwards from Cape Town, the coast becomes progressively more wooded and subtropical, the ocean warms, the rains fall year-round, and the forests host an ever-greater diversity of birds. The region from Mossel Bay to the Tsitsikamma area is a rather paradisiacal stretch of coastal belt that, thanks to its pleasing climate, secluded beaches and still extensive tracts of canopy forest, has become a favoured recreational destination aptly known as the Garden Route. Further inland, a dramatic, fynbos-clad barrier of mountains gives way to the arid expanse of the Karoo, transecting a remarkable diversity of habitats that offer rewarding birding.

> ### TOP BIRDS
> Forest Buzzard, Crowned Eagle, Knysna Lourie, Emerald Cuckoo, Half-collared Kingfisher, Narina Trogon, Knysna Woodpecker, Short-toed Rock Thrush, Chorister Robin, Cape Rockjumper, Knysna Warbler, Olive Bush Shrike, African Rock Pipit, Southern Tchagra, Protea Canary.

Birders will find it easy to escape the droves of holiday-makers, who tend to keep close to the major centres of Knysna and Plettenberg Bay. You can potter about the restful coastal lake system of the Wilder-

Knysna Lourie

The extensive forests at Nature's Valley provide some of the best birding along the Garden Route.

cross-section of Karoo birding. Besides being a very worthwhile destination in itself, it serves as an excellent staging post en route from Cape Town to Johannesburg.

THE WILDERNESS REGION

At the western end of the Garden Route, the Wilderness National Park encloses a system of reed-fringed coastal lakes which are sandwiched between a beach and an escarpment cloaked with lush coastal forest. A number of tranquil paths lead through the forest, offering easy and pleasant access to an excellent selection of forest birds, including **Knysna Lourie** (p.125*). The park's rest camp is at Ebb and Flow (① on site map, p.120), a 1-km drive north of the N2 national road (turn inland at the signs just east of Wilderness village). The hutted camp and campsite, particularly at Ebb and Flow North (adjacent to Ebb and Flow south), offers good birding (including nightly **Wood Owls**) and is a rewarding place to base your explorations into the surrounding forests.

Leading from Ebb and Flow are four trails, all named after local kingfishers. Good forest birding is to be had along the Giant Kingfisher Trail, which begins at the northern end of the Ebb and Flow North campsite. It runs alongside the eastern bank of the Touw River, ascending its forested valley and ultimately reaching a waterfall at ②, 3.5 km from the campsite. The most conspicuous species in the forest are usually **Bar-throated Apalis** and **Green-backed Bleating Warbler**. **Terrestrial Bulbul** and **Chorister Robin** (p.125*) lurk in the lower strata (see box opposite), while common species of the mid-canopy are **Cape Batis**,

ness National Park, or take refuge in the lush forests protected here and at Nature's Valley in the Tsitsikamma National Park, where a good diversity of forest specials is on offer. Sadly, only a single elephant still survives in the forests, a remnant of the great herds that once roamed the area.

This verdant coastal region is isolated from the interior of the country by the Outeniqua and Tsitsikamma mountains. In their rain shadow lies the dry Little Karoo, a region made famous and prosperous by its ostrich farmers, who supplied the demands of an ephemeral ladies' fashion a century ago. North of the Little Karoo lies another great fold in the landscape, the Swartberg Mountains, traversed by the ambitious bends of the Swartberg Pass. Here, notably, six of the fynbos endemic birds may readily be found close to the road. The Swartberg is also the last outpost of moist landscape before the vast and arid Great Karoo, which covers much of central South Africa.

Contrary to the preconceptions of many, the Karoo presents a varied and exciting landscape offering excellent birding (see also pp.74 and 86). The mountainous Karoo National Park, near the regional centre of Beaufort West (five hours' drive on the N1 from Cape Town), provides a representative

Dusky Flycatcher, Sombre Bulbul, Yellow-throated Warbler, Blue-mantled Flycatcher, Olive Woodpecker and, less conspicuously, Olive Bush Shrike. It is especially important to familiarize yourself with the calls of otherwise cryptic canopy species such as Black-headed Oriole, Narina Trogon (p.125*), Grey Cuckooshrike, Scaly-throated Honeyguide, Black-bellied Starling, Knysna Lourie (p.125*), and Red-billed Woodhoopoe. The latter two birds often venture into the rest-camp edges.

Look out for Red-necked Francolin feeding cautiously in open areas near the forest edge, especially in the early morning and in the evening. Listen for the deep hoot of Buff-spotted Flufftail at night (see p.20). Crowned Eagle and African Goshawk may be seen overhead here, as indeed in any of the Garden Route forests. Cinnamon (the most common) and Tambourine Doves are birds of the forest floor and are most often seen when flushed.

The lakes offer good birding.

Also providing good access to these forest species is the Half-collared Kingfisher Trail, running along the other side of the Touw River for 2.5 km. It offers a good view of the river edges, and thus better access to its scarce and inconspicuous namesake, which is resident along its length. The reeds along the Touw River at the Ebb and Flow campsite edges have hosted Great Reed Warbler, which may be a scarce visitor here despite being almost unknown in the Cape.

In the Wilderness system, the best lakes for birding are Langvlei and Rondevlei, largely because both have well-positioned hides (at ③ and ④ respectively, see map overleaf) which are accessed along boardwalks

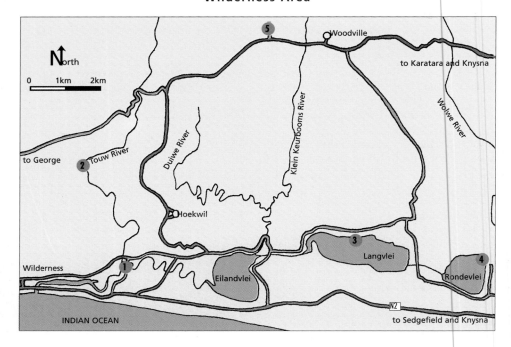

cut through dense reedbeds. These hides, especially the one at Rondevlei, are excellent places to search for stubborn skulkers such as **Red-chested Flufftail** and **African Rail**. The best way to see these birds is to lure them across the gap in the reeds formed by the boardwalk. **Baillon's Crake** also occurs here, but is less likely to be enticed into view and is best searched for at the reed edges in the early morning. Rallids aside, the lakes offer a pleasant selection of more conspicuous species, including **Yellow-billed Egret**, **Purple Heron**, **African Fish Eagle**, **Osprey** (summer), **African Marsh Harrier**, **Malachite Kingfisher** and **Cape Reed** and **African Sedge Warblers**.

Excellent forest birding (possibly even superior to that at Wilderness) may also be enjoyed in Woodville Forest, just to the north of the lakes (see map, above; follow the signs to the 'Big Tree' at ⑤). A few minute's walk from the parking area leads you to the aptly named tree – a gargantuan Outeniqua Yellowwood (*Podocarpus falcatus*) – from which a 2-km footpath loops gently through the forest. All the birds listed for the Wilderness forests occur here. Furthermore, Woodville is probably a better site for **Starred Robin**, **Knysna Woodpecker** (albeit scarce; p.72*) and, in streamside undergrowth and scrubby forest edge habitats, **Knysna Warbler** (p.32*). Secondary growth at forest edges – such as

Red-chested Flufftail: reed dweller.

that along the short section of road leading to the parking area – is worth checking for **Forest Canary**, **Swee Waxbill** and **Greater Double-collared Sunbird**.

Another excellent site for **Knysna Warbler** (p.125*) is Victoria Bay, situated between Wilderness and George. Take the signposted turn-off south from the N2 and follow this winding road all the way down to the beach, where there is a gate and parking area. **Knysna Warbler** is common in the undergrowth of the adjacent dense coastal thicket (for example, near the boardwalk on the left of the parking area).

Forest Buzzard occurs in forests and plantations throughout the Garden Route, and regularly perches on roadside telephone poles along the N2 from Wilderness through Knysna and to Nature's Valley. In summer, it is joined by **Steppe Buzzard**, posing an identification challenge.

KNYSNA AND PLETTENBERG BAY

The town of Knysna, immortalized in several bird names and perched on the edge of the Knysna Lagoon, lies midway between Wilderness and Nature's Valley. The N2 crosses the northern end of the lagoon, providing a vantage point from which to scan for the sometimes large numbers of waterbirds. You may wish to visit the famous Knysna Heads, guarding the lagoon mouth, as much for their scenic attraction as for **Knysna Warbler**, which is resident in the coastal undergrowth between the parking area and the sea. The signposted turn-off to the Heads is in the commercial centre of Knysna, 8.5 km east of the bridge over the lagoon. Check the nearby Woodbourne Lagoon for an excellent selection of waders and other waterbirds. Typical forest and forest edge birding is available close to town, and a wide variety of options is available, including the Pledge Nature Reserve and Diepwalle Forest Walk. Consult Knysna Tourism for further information on the forest walks.

The popular holiday town of Plettenberg Bay offers a pleasant selection of waterbirds on the Keurbooms River estuary and Bitou River. The nearby Robberg Peninsula holds a number of fynbos specials. Although this area does not have extensive forests, enquire about the Piesang River Nature Trail on the edge of town. **Southern Tchagra** may be found in the scrubby coastal areas.

NATURE'S VALLEY

At the eastern end of the Garden Route lies the Tsitsikamma National Park, incorporating a stretch of coastline altogether more rugged than that of the Wilderness region. Excellent forest birding – arguably the best along the Garden Route – is easily accessible in the park's western De Vasselot section. Here, large tracts of pristine forest surround the strikingly picturesque coastal village of Nature's Valley. To reach the latter, follow the N2 national road east of Plettenberg Bay for 28 km, and turn south onto the R102 just before the toll gate. The road winds down the densely forested Grootrivier Pass for 12 km before reaching the coast and Nature's Valley village. Continue a few hundred metres past the village and park at the De Vasselot campsite, where you need to obtain a permit at the National Parks office. The Grootrivier Trail, looping a gentle 4.5 km around the forested Groot River lagoon,

The Groot River at Nature's Valley.

Above: Crowned Eagle, a forest inhabitant.
Below: Taller streamside vegetation along Swartberg Pass supports Victorin's Warblers.

offers convenient access to fine forest birding. The marked trail begins on the eastern side of the Groot River bridge, a minute's walk further along the tar road. **African Finfoot** occurs scarcely along this stretch of river: from the bridge, carefully scan overhanging vegetation at the river edges, and check the series of pools between the river and the campsite. **Half-collared Kingfisher** is also occasionally seen here. All the forest species listed on p.118 for Wilderness National Park occur commonly close to this bridge. Additionally, at Nature's Valley there is an increased likelihood of encountering **Crowned Eagle**, **Little Sparrowhawk**, **Emerald Cuckoo** (summer), **Scaly-throated Honeyguide**, **Knysna Woodpecker** and **Starred Robin**. At night, you are likely to hear **Wood Owl** and **Buff-spotted Flufftail**, although the latter is notoriously difficult to see (see p.20 for some tips).

Beyond De Vasselot and east of Nature's Valley, the R102 winds up the spectacular Bloukrans Pass, which provides a good vantage point from which to scan for **Crowned Eagle**. **Victorin's Warbler** (p.73*) occurs in the dense fynbos at the roadside. Shortly before the R102 rejoins the N2 national road there is a sawmill, adjacent to the road, where **Black-winged Plover** is occasionally seen.

SWARTBERG PASS

For over a century, the 24-km length of the Swartberg Pass has connected the Little Karoo ostrich capital of Oudtshoorn to the placid Great Karoo town of Prince Albert. As you head north from Oudtshoorn and the road begins to climb, there is a sudden switch from arid scrub and ravine-side thicket to moist, mist-wreathed mountain fynbos. The pass crests the Swartberg range at an altitude of 1 436 m before dropping precipitously through a series of dramatic switchbacks, ingeniously supported by dry-stone walls. Becoming progressively more arid, the road emerges into the Karoo proper through a kloof presided over by agonizingly contorted rock layers folded by massive early geological upheavals.

To reach the pass, take the R328 from Oudtshoorn, and follow the signs. You might like to spend a moment investigating the dry hillside scrub in the vicinity of the Cango Mountain Resort turn-off (24 km north of Oudtshoorn) where, among others, **Layard's Titbabbler** and **Fairy Flycatcher** may be found. By the time the pass proper begins (where the road surface changes to gravel, 43 km north of Oudtshoorn), the altitude has rapidly transformed the parched scrub into moist, dense mountain fynbos. The distances given are measured from the end of the tarred road at this point; those in brackets are measured from the beginning of the tarred road on the other side of the pass, 2 km from Prince Albert. At 0.5 (25.5) km past the transition to gravel, a stream flanked by taller vegetation passes under the road.

Geological contortions along Swartberg Pass.

This is an excellent site for **Protea Canary** (p.57*); look particularly in the tall streamside growth, and in the adjacent stands of Waboom (*Protea nitida*, a tall, greyish-leafed protea). **Protea Canary** is in fact common in taller vegetation on both sides of the pass. More conspicuous species to be found in this vicinity are **Cape Sugarbird** (p.33*), **Orange-breasted Sunbird** (p.33*), **Neddicky, Cape Bulbul, Grassbird** and **Malachite Sunbird. Victorin's Warbler** is very common in the impenetrable undergrowth, supported by seeps, along the entire ascent of the pass (from here to the summit); look for them at the streams at 3.8 (22.2), 5.0 (21.0) and 7.3 (18.7) km. As the road approaches its highest altitude, the terrain becomes ever-rockier and cooler. The rocky ridges to the east and west of the pass's summit at 9.2 (16.8) km ('Die Top') are well worth a walk in quest of **Ground Woodpecker** (p.105*), **Cape Rockjumper** (p.73*), **Sentinel** and **Cape Rock Thrushes** and **Cape Siskin** (p.33*)

North of the summit, the landscape is noticeably drier, revealing spectacular geological contortions. Keep an eye out for classic mountain raptors such as **Black** and **Booted Eagles, Jackal Buzzard** and **Rock Kestrel**. The road descends through progressively more arid country and ultimately joins a river in a dry gorge, at 22.2 (3.8) km. The hillside scrub flanking it hosts such typically kloof-loving Karoo species as **Fairy Flycatcher, Layard's Titbabbler** and, usually flying overhead, **Pale-winged Starling**. At 25.7 (0.3) km, a picnic site lies on the left hand side alongside a band of fearsomely thorned acacia trees. This thicket offers **Southern Tchagra, Pririt Batis** (p.85*), **Namaqua Warbler** (p.85*) and **Red-billed Firefinch**, the latter near the western limit of its range. From the junction at 26.0 (0.0) km, the roads runs for 2 km through a broad valley to the sleepy town of Prince Albert and the Great Karoo beyond.

KAROO NATIONAL PARK

The park protects an exceptionally fine tract of mountainous Karoo landscape near the town of Beaufort West, and is well-stocked with game – including Black Rhino (*Diceros bicornis*), Black Wildebeest (*Connochaetes gnou*), Gemsbok (*Oryx gazella*) and Cape Mountain Zebra (*Equus zebra zebra*). It also boasts an excellent selection of Karoo specials, providing you with easy access to three tricky rock-loving species, namely **Cinnamon-breasted Warbler** (p.85*), **African Rock Pipit** (p.125*) and **Short-toed Rock Thrush**. Another of the park's attractions is the opportunity to join guided night-drives and go spotlighting for some exciting Karoo mammals, among them Aardvark, Caracal and Aardwolf (see box, p.124). **Cape Eagle Owl** (p.105*) may also, very occasionally, be seen on these night drives.

The entrance to the national park is on the N1 national road, 5 km south of Beaufort West. A tarred road leads to the park's headquarters and rest camp (①on map overleaf), where **Mountain Chat, Red-eyed Bulbul** and **Cape Bunting** are tame and conspicuous. **Black Eagle** regularly passes overhead. Take an amble around the nearby campsite (②), set in dense acacia thicket, as it offers some of the best birding in the park. **Namaqua Warbler** (p.85*), **Southern Tchagra, Acacia Pied Barbet, Cardinal Woodpecker, Dusky Sunbird, Pririt Batis, Titbabbler** and **Fairy Flycatcher** are all vocal but inconspicuous thicket dwellers. Rather more obvious are all three South African **mousebird** species.

Karoo National Park

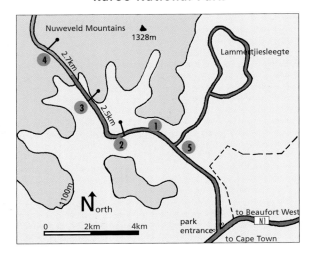

both **Cinnamon-breasted Warbler** and **African Rock Pipit**, neither of which is likely to be seen without staying very alert to their calls, which drift across from the cliff faces. The pipit also occurs on the rocky hillocks of the plateau itself, alongside the similarly rock-loving **Long-billed Pipit**. Other mountain species that are typical of the cliffs along the Klipspringer Pass are **Black** and **Booted Eagles**, **Ground Woodpecker** (p.105*) and **Pale-winged Starling**. Continue on to the plateau, where the road moves into more open country and **Sickle-winged Chat** and **Chat Flycatcher** occur.

Klipspringer Pass drive, which winds up the escarpment of the plateau behind the rest camp, provides access to the three specials of rocky country. Check the slopes at the base of the pass (in the vicinity of ③) for **Layard's Titbabbler**, **African Rock Pipit** (opposite) and **Short-toed Rock Thrush**. The latter occurs scarcely but regularly along the length of the meandering road up to the pass; the birds here are of the central dryland subspecies *pretoriae*, which has been regarded as a full species by some authors (see p.13). The vicinity of the fenced lookout point at the summit of the pass at ④ is a good site for

View from Klipspringer Pass.

NOCTURNAL MAMMALS OF THE KAROO

Night-drives in the Karoo can be every bit as exciting as those in the renowned game reserves of the savanna regions. There may be no lions, but for many the experience of seeing such unique and bizarre mammals as **Aardvark** (*Orycteropus afer*, below), Caracal (*Felis caracal*), Porcupine (*Hytrix africaeaustralis*) and Aardwolf (*Proteles cristata*) will be even more memorable. More common predators are Bat-eared Fox (*Otocyon megalotis*), African Wild Cat (*Felis sylvestris*), Cape Fox (*Vulpes chama*) and Black-backed Jackal (*Canis mesomelas*). You may be lucky enough to see one of these species foraging at dawn or dusk.

Take a gentle drive along the 13-km Lammertjiesleegte circular road (starting at ⑤), for superb Karoo plains birding as well as the park's best diversity of game. Scan open patches for ground birds like **Ostrich**, **Ludwig's Bustard** (p.105*), **Karoo Korhaan**, **Double-banded Courser** and **Namaqua Sandgrouse**, and check all small passerines flushing up from the road-side: **Karoo**, **Spike-heeled**, **Karoo Long-billed** and **Sabota Larks** are all reasonably common, along with the more conspicuous **Karoo**, **Tractrac** (see p.78) and **Anteating Chats**. Also look for **Karoo Eremomela** (p.85*). **Black-headed Canary** (p.105*) and **Lark-like Bunting** are especially obvious after rains, when **Black-eared Finchlark** (p.96*) also occasionally enters the area.

SELECT SPECIALS

Narina Trogon

Named after the Khoikhoi mistress of the 18th-century French explorer and ornithologist François le Vaillant, the Narina Trogon has a quite unfounded reputation for elusiveness. The reality is that this bird is not particularly shy, and is easily deceived by imitations of its call. Note, however, that trogons invariably sound much further away than they are – you may find yourself patiently trying to lure in a seemingly distant bird that is in fact only two trees away from you. They swoop in quietly to perch in the canopy, usually facing away from you, thus turning a cryptic green against the forest canopy. Look out for the beautiful patches of electric blue skin on their faces that puff out as they call.

Knysna Lourie

This charismatic endemic makes a worthy icon of the Garden Route. It is usually first observed as a flash of red gliding through the green foliage, and is easy to locate by its hoarse, repetitive and very loud call, rising to a crescendo – one of the most evocative sounds of the mist-wreathed canopy of South Africa's afromontane forests. The Knysna Lourie is perhaps less shy than other forest louries and appears insatiably curious. Many a lunch stop will be enlivened by the discovery of a lourie peering at you through the overhead leaves.

Chorister Robin

The Afrikaans name for this species, 'Lawaaimaker', is rather less generous than the English! It means 'racket-maker', but however melodious one considers it, the Chorister Robin is a very fine-looking endemic. In the Western Cape, its range only extends eastwards from Mossel Bay, and it occurs in all the Garden Route forests described here. It is perhaps typical of many forest species in that it is brightly coloured and vocal, but often infuriatingly skulking. It calls mostly at dawn, often from the forest canopy, which may make it difficult to locate. Beware of its remarkable powers of mimicry, which may be to blame if you are baffled by Crowned Eagles or African Goshawks calling from dense forest understorey!

African Rock Pipit

This poorly-known endemic is characteristic of South Africa's mountainous interior. It is localized and largely inaccessible in the region covered by this book, with the notable exception of the Karoo National Park (opposite); see also Bergenaars Pass (p.112). African Rock Pipit is curiously inconspicuous and best detected by call – a rather unpipit-like descending whistle. The yellow edging on the bird's folded wing, rather optimistically exaggerated in some field guides, is not a good field character. Rather, concentrate on its very distinctive call, plain plumage and relatively conspicuous eye-stripe to distinguish it from the Long-billed Pipit, which favours similarly rocky landscapes.

CHECKLISTS AND INDEX OF THE BIRDS OF WESTERN SOUTH AFRICA

This is a complete checklist of the birds of South Africa's Western and Northern Cape Provinces. It follows the standard field-guide sequence and includes, in column order from left to right: the common name with, where applicable; the proposed new name in brackets (where the name change involves only a minor style change, the new name is given); a selection of the page references to fit (sometimes the names are abbreviated); whether the bird is endemic (**E**) or near-endemic (**N**) to either southern (**sA**) or South Africa (**SA**), see p.5; the seasonality of records (**Y: A** = all year, **S** = summer, **W** = winter); and whether the species has been recorded in the Western (**WC**) or Northern (**NC**) Cape Province. It also gives the status of each species in each region covered by this book. The regions are as follows: **1** = Cape Peninsula; **2** = Seabirding; **3** = West Coast; **4** = Overberg and South Coast; **5** = Tanqua Karoo Loop; **6** = Bushmanland; **7** = Namaqualand; **8a** = Western Kalahari (west of 22° E longtitude); **8b** = Eastern Kalahari; **9** = Garden Route and Interior. In each of these regions, **1** = Likely to be seen in 2–3 days birding in the region specified, using this book as a guide; **2** = Uncommon or local: may be seen while birding in the region; **3** = rare or very localized: unlikely to be seen on a visit to the region.

Name (Proposed new name) Scientific name Page references	sA	SA	Y	WC	NC	1	2	3	4	5	6	7	8A	8B	9
Ostrich (Common Ostrich) *Struthio camelus* 24,47,65,99,125			A	1	1	1		1	1			2	1	1	1
King Penguin *Aptenodytes patagonicus*			S	1				3							
Gentoo Penguin *Pygoscelis papua*			W	1				3							
African (Jackass) Penguin *Spheniscus demersus* 8,14,24,30-2,55	N		A	1	1	1	1	1	1			2			2
Rock-hopper Penguin *Eudyptes chrysocome*			A	1	1	3		3				3			3
Macaroni Penguin *Eudyptes chrysolophus*			S	1		3		3							
Great Crested Grebe *Podiceps cristatus* 29,56,66			A	1	1			1	1	1		3		1	1
Black-necked Grebe *Podiceps nigricollis* 27-8,53,89,115			A	1	1	1		2	1	2	2	2	3	2	2
Dabchick (Little Grebe) *Tachybaptus ruficollis* 42			A	1	1	1		1	2	1	2	2	2	2	1
Royal Albatross *Diomedea epomophora* 35-6			W	1			3								
Wandering Albatross *Diomedea exulans* 35-6,39,40			W	1	1		2								
Laysan Albatross *Diomedea immutabilis* 37			S	1			3								
Shy Albatross *Diomedea cauta* 35-6,39			A	1	1	1	1	2	2			3			2
Black-browed Albatross *Diomedea melanophris* 35-6,39			A	1	1	1	1	2	2			3			2
Grey-headed Albatross *Diomedea chrysostoma* 35-6			W	1			3								
Buller's Albatross *Diomedea bulleri* 36			W	1			3								
Yellow-nosed Albatross *Diomedea chlororhynchos* 35-7,39			A	1	1	2	1	3	1			3			2
Dark-mantled Sooty Albatross (Sooty Albatross) *Phoebetria fusca* 36-7			W	1		3	3	3							
Light-mantled Sooty Albatross *Phoebetria palpebrata* 37			W	1			3								
Southern Giant Petrel *Macronectes giganteus* 35-6,39,40			A	1	1	2	1	2	2			3			2
Northern Giant Petrel *Macronectes halli* 35-6,39			A	1	1	2	1	2	2			3			2
Antarctic Fulmar (Southern Fulmar) *Fulmarus glacialoides* 35-6			W	1	1	3	3								
Antarctic Petrel *Thalassoica antarctica* 37			W	1			3								
Pintado Petrel *Daption capense* 35-6			A	1	1	3	1	3	3						3
Great-winged Petrel *Pterodroma macroptera* 35-6			S	1	3	3	2								
Soft-plumaged Petrel *Pterodroma mollis* 35-6,39			W	1	3	3	2								
White-headed Petrel *Pterodroma lessonii* 37			W	1			3								
Atlantic Petrel *Pterodroma incerta* 37			W	1			3								
Kerguelen Petrel *Lugensa brevirostris* 37			W	1			3								
Blue Petrel *Halobaena caerulea* 37			W	1			3								
Broad-billed Prion (Antarctic Prion) *Pachyptila vittata* 36,39			W	1	1	2	1								
Slender-billed Prion *Pachyptila belcheri* 37			W	1			3								
Fairy Prion *Pachyptila turtur* 37			W	1			3								
White-chinned Petrel *Procellaria aequinoctialis* 35-6,39,52			A	1	1	1	1	1	1			1			2
Spectacled Petrel *Procellaria conspicillata* 35-6,40			A	1	1		2								
Grey Petrel *Procellaria cinerea* 37			W	1			3								
Cory's Shearwater *Calonectris diomedea* 35-6,39			S	1	1	2	1	2							
Great Shearwater *Puffinus gravis* 35-6,39			A	1	1	3	1								
Flesh-footed Shearwater *Puffinus carneipes* 35-6			W	1		3	3								
Sooty Shearwater *Puffinus griseus* 35-6,39			A	1	1	1	1	1	1			2			2
Manx Shearwater *Puffinus puffinus* 35-36			S	1	1	3	2								
Little Shearwater *Puffinus assimilis* 37			W	1			3								
European Stormpetrel *Hydrobates pelagicus* 35-6			S	1	1		3								
Leach's Stormpetrel *Oceanodroma leucorhoa* 35-6,40			S	1	1		2		3						
Wilson's Stormpetrel *Oceanites oceanicus* 35-6,39			A	1	1	2	1	3	3			3			3
White-bellied Stormpetrel *Fregetta grallaria* 37			A	1			3								
Black-bellied Stormpetrel *Fregetta tropica* 35-6			A	1			2								
Red-billed Tropicbird *Phaethon aethereus*			A	1		3									
Red-tailed Tropicbird *Phaethon rubricauda*			A	1		3		3							
White-tailed Tropicbird *Phaethon lepturus*			A	1				3							
White Pelican (Great Wh. Pel.) *Pelecanus onocrotalus* 28-9,42,53-4,56			A	1	1	1		1	1			1			3
Pink-backed Pelican *Pelecanus rufescens*			A		1									3	
Red-footed Booby *Sula sula*			S	1		3									
Cape Gannet *Morus capensis* 8,35-6,39,41,52,55	N		A	1	1	1	1	1	1			1			1
Australian Gannet (Australasian Gannet) *Morus serrator* 55			A	1			3								
White-breasted Cormorant (Great Corm.) *Phal. carbo* 22,24,29,52,62			A	1	1	1	1	1	1	1	1	1	1	1	1
Cape Cormorant *Phalacrocorax capensis* 8,22,24-5,29,31,39,52,62	N		A	1	1	1	1	1	1			1			1
Bank Cormorant *Phalacrocorax neglectus* 8,14,21-2,31,46,62,101	E		A	1	1	1	2	1	1			2			1
Reed Cormorant *Phalacrocorax africanus*			A	1	1	1		1	1	1	1	1	1	1	1
Crowned Cormorant *Phal. coronatus* 8,22,24-5,31,43,46,52,62,101	E		A	1	1	1	1	1	1			1			3
Darter (African Darter) *Anhinga melanogaster* 29,42,94			A	1	1	1		1	1	1		1		1	1

	sA	SA	Y	WC	NC	1	2	3	4	5	6	7	8A	8B	9
Greater Frigatebird *Fregata minor*			A	1				3							
Grey Heron *Ardea cinerea*			A	1	1	1	1	1	1	1	1	1	1	1	1
Black-headed Heron *Ardea melanocephala*			A	1	1	1		1	1	1	1	1	1	1	1
Goliath Heron *Ardea goliath*			A	1	1		3	3	3	1	1	2	1	3	
Purple Heron *Ardea purpurea* 29,54,56,120			A	1	1		1	2	1	3	3	3	1	1	
Great White Egret (Great Egret) *Casmerodius albus*			A	1	1	3		3	3	3			3	2	2
Little Egret *Egretta garzetta* 24,42,54			A	1	1	1		1	1	1	2	1	3	1	1
Yellow-billed Egret (Intermediate Egret) *Mesophoyx intermedia* 42,120			A	1	1	2		1	1	1	3	3	3	2	1
Black Egret *Egretta ardesiaca*			A	1	1	3		3						2	
Little Blue Heron *Egretta caerulea* 50,54			A	1				3							
Cattle Egret *Bubulcus ibis*			A	1	1	1		1	1	1	2	2	1	1	1
Squacco Heron *Ardeola ralloides*			A	1	1	3		3	3	3		3	3	1	3
Green-backed Heron *Butorides striatus*			A	1	1							3		3	3
Rufous-bellied Heron *Ardeola rufiventris*			A	1									3		
Black-crowned Night Heron *Nycticorax nycticorax* 84			A	1	1	1		1	1	1		2	3	2	1
White-backed Night Heron *Gorsachius leuconotus*			A	1											3
Little Bittern *Ixobrychus minutus* 29,56,83,84			A	1	1	2		2	2	1	1	3	3	2	2
Dwarf Bittern *Ixobrychus sturmii*			S	1	1	3							3		3
Bittern (Great Bittern) *Botaurus stellaris*			Locally	extinct											
Hamerkop *Scopus umbretta*			A	1	1	3		2	1	2	2	1	1	1	1
White Stork *Ciconia ciconia* 64			S	1	1	3		3	1	2	3	3	2	1	1
Black Stork *Ciconia nigra* 112			A	1	1	3		2	3	3	2	3	3	2	1
Abdim's Stork *Ciconia abdimii* 111			S	1		3			3				2	2	
Woolley-necked Stork *Ciconia episcopus*			A	1									3		
Open-billed Stork (African Openbill) *Anastomus lamelligerus*			A	1				3							
Saddle-billed Stork *Ephippiorhynchus senegalensis*			A	1										3	
Marabou Stork *Leptoptilos crumeniferus*			A	1	1	3			3		3	3	3	3	
Yellow-billed Stork *Mycteria ibis*			A	1	1	3		3						2	3
Sacred Ibis *Threskiornis aethiopicus* 26			A	1	1	1		1	1	1	1	2	1	1	1
Bald Ibis (Southern Bald Ibis) *Geronticus calvus*	E	E	Locally	extinct											
Glossy Ibis *Plegadis falcinellus* 42			A	1	1	1		1	2	1	3		3	2	1
Hadeda Ibis *Bostrychia hagedash* 26			A	1	1	1		1	1	3	2	3	1	1	1
African Spoonbill *Platalea alba* 26,50,54			A	1	1	1		1	1	1	2	3	1	1	1
Greater Flamingo *Phoenicopterus ruber* 28,53-6,77,84,101,104,115			A	1	1	1		1	2	1	1	3	1	1	3
Lesser Flamingo *Phoenicopterus minor* 53-4,56,83-4,115			A	1	1	3		1	2	1	2	2	3	1	3
Mute Swan *Cygnus olor INTRODUCED* 31			A	1					3						
White-faced Duck *Dendrocygna viduata* 84			A	1	1	3		3		2	3	3	3	1	3
Fulvous Duck *Dendrocygna bicolor*			A	1	1	3		3	3	3				3	2
White-backed Duck *Thalassornis leuconotus* 22,42,69,83-4			A	1	1	2		1	2	1				3	3
Egyptian Goose *Alopochen aegyptiacus*			A	1	1	1		1	1	1	1	1	1	1	1
South African Shelduck *Tadorna cana* 28-9,49,52,54-6,77,89,101,104	E		A	1	1	1		1	1	1	1	1	1	1	1
Yellow-billed Duck *Anas undulata* 27,42			A	1	1	1		1	1	1	1	1	1	1	1
African Black Duck *Anas sparsa* 62,69,83-4,113			A	1	1	2		3	2	2	2	2	2	2	2
Mallard *Anas platyrhynchos INTRODUCED* 31			A	1	1	1		2	2	2					
Cape Teal *Anas capensis* 27-8,53,55-6,101,104			A	1	1	1		1	1	1	1	1	1	1	1
Hottentot Teal *Anas hottentota* 28			A	1	1	3		3	3	3	3	3	3	2	
Red-billed Teal *Anas erythrorhyncha* 27,55			A	1	1	1		1	1	1	2	3	2	1	1
Garganey *Anas querquedula*			S	1		3									
Northern Shoveller *Anas clypeata*			S	1											
Cape Shoveller *Anas smithii* 27,42	N		A	1	1	1		1	1	1	2	2	3	1	1
Southern Pochard *Netta erythrophthalma* 27-8,84			A	1	1	1		2	2	1	3	3	3	1	2
Pygmy Goose *Nettapus auritus*			A	1		3									
Knob-billed Duck *Sarkidiornis melanotos*			A	1	1	3		3	3	3				3	
Spur-winged Goose *Plectropterus gambensis*			A	1	1	2		1	1	1	2	3	3	1	1
Maccoa Duck *Oxyura maccoa* 22,27-8,84,101			A	1	1	1		3	3	1	2	3	2	3	2
Secretarybird *Sagittarius serpentarius* 45,64-5,69,109			A	1	1	3		2	2	3	2	2	1	1	2
Bearded Vulture *Gypaetus barbatus*			Locally	extinct											
Egyptian Vulture *Neophron percnopterus*			Locally	extinct											
Hooded Vulture *Necrosyrtes monachus*			A	1									3		
Cape Vulture *Gyps coprotheres* 58-9,64-5,72	E		A	1	1	3		3					3	3	3
Rüppell's Vulture *Gyps rüpellii*			A	1				3							
White-backed Vulture (African Wh-bkd Vul.) *Gyps africanus* 109,114			A	1	1							3		1	1
Lappet-faced Vulture *Torgos tracheliotus* 109,112			A	1										1	2
White-headed Vulture *Trigonoceps occipitalis* 109			A	1										2	
Black Kite *Milvus migrans*			S	1								3		3	3
Yellow-billed Kite *Milvus parasitus* 44			S	1	1	2		1	1	3	3	3	3	2	3
Black-shouldered Kite *Elanus caeruleus* 43,47			A	1	1	2		1	1	1	1	1	1	1	1
Cuckoo Hawk (African Baza) *Aviceda cuculoides* 70			A	1				3							3
Honey Buzzard (European Honey Buzzard) *Pernis apivorus* 20			S	1		2									
Black Eagle (Verreaux's Ea.) *Aq. verreauxii*18,52,61,89,99,103,112,123			A	1	1	2		1	1	2	1	2	1	2	1
Tawny Eagle *Aquila rapax* 109			A	1										1	2	3
Steppe Eagle *Aquila nipalensis* 109			S	1	1	1			3					2	3
Lesser Spotted Eagle *Aquila pomarina* 109			S	1										3	3
Wahlberg's Eagle *Aquila wahlbergi*			S	1										3	3	3
Booted Eagle *Hieraaetus pennatus* 56,80,89,99,103,123-4			A	1	1	2		1	1	1	1	1	1	2	1
African Hawk Eagle *Hieraaetus spilogaster*			A	1										3	3
Long-crested Eagle *Lophaetus occipitalis*			A	1		3									
Martial Eagle *Polemaetus bellicosus* 61,64,69,89,109			A	1	1	3		3	2	2	1	1	1	2	1
Crowned Eagle (African Cr. Ea.) *Stephanoaetus coronatus* 69,117,122			A	1				2							2

	sA	SA	Y	WC	NC	1	2	3	4	5	6	7	8A	8B	9
Brown Snake Eagle *Circaetus cinereus* 109			A	1	1			3	3				2	3	
Black-breasted Snake-Eagle *Circaetus pectoralis* 89,101,109			A	1	1			3	3	3	1	1	2	2	3
Bateleur *Terathopius ecaudatus* 106-7,109			A	1	1			3				1	1	3	
Palm-nut Vulture *Gypohierax angolensis*			A	1	1			3	3				3		3
African Fish Eagle *Haliaeetus vocifer* 26,56,84,94,101,120			A	1	1	2		2	2	1	1	1	1	1	1
Steppe Buzzard (Common Buzzard) *Buteo buteo* 20,43,54			S	1	1	1		1	1	1	2	1	1	1	1
Forest Buzzard *Buteo trizonatus* 8,20,26,70,117,121	E	E	A	1		2			1						1
Jackal Buzzard *Buteo rufofuscus* 18,24,45,52,54,61-2,89,99,101,123	E		A	1	1	2		1	1	1	1	1	2	2	1
Red-breasted Sparrowhawk *Accipiter rufiventris* 17-8,20,61			A	1	1	1		3	2	2	3				1
Ovambo Sparrowhawk (Ovampo Sparrowhawk) *Accipiter ovampensis*			A	1										3	
Little Sparrowhawk *Accipiter minullus* 122			A	1				3							2
Black Sparrowhawk (Black Goshawk) *Acc. melanoleucus* 17-8,26,65,70			A	1	1	2		3	1	3					1
Little Banded Goshawk (Shikra) *Accipiter badius* 113			A	1									3	2	
African Goshawk *Accipiter tachiro* 17-8,20,65,70,119			A	1	1			3	1	2					1
Gabar Goshawk *Micronisus gabar* 109			A	1	1	3			3		3	3	1	1	2
Pale Chanting Goshawk *Melierax canorus* 76-8,89,101,109	N		A	1	1			3	3	1	1	1	1	1	1
European Marsh Harrier (Eurasian Marsh Harrier) *Circus aeruginosus*			S	1									3		
African Marsh Harrier *Circus ranivorus* 26,29,43,49,56,64,101,120			A	1	1	1		1	1	2		2		2	1
Montagu's Harrier *Circus pygargus* 109			S	1	1			3					3	3	
Pallid Harrier *Circus macrourus*			S	1		3		3					3	3	
Black Harrier *Circus maurus* 41,47,49,57,61,63-5,69,70,81,89,109	E		A	1	1	3		1	1	2	1	3	2	2	2
Gymnogene (African Gymnogene) *Polyboroides typus* 65			A	1	1	3		2	3	3	3	2	2	3	1
Osprey *Pandion haliaetus* 49,120			S	1	1	3		1	2				3	3	1
Peregrine Falcon *Falco peregrinus* 18,21-3,49,61,83,112			A	1	1	1		2	2	2	3	2	2	3	2
Lanner Falcon *Falco biarmicus* 18,52,89,99,101,109			A	1	2			1	2	2	1	1	1	2	1
Sooty Falcon *Falco concolor*			S	1									3		
Hobby Falcon (Eurasian Hobby) *Falco subbuteo*			S	1	1	3		3	3				3	3	3
African Hobby *Falco cuvierii*			A	1									3		
Eleonora's Falcon *Falco eleonorae*			S	1				3							
Red-necked Falcon *Falco chicquera* 106-7,109-10			A	1							3		1	3	
Western Red-footed Kestrel (Red-footed Falcon) *Falco vespertinus*			S	1									3	3	
Eastern Red-footed Kestrel (Amur Falcon) *Falco amurensis* 69			S	1	1			2		3			3	3	3
Rock Kestrel *Falco tinnunculus* 18,21,52,61,80,123			A	1	1			1	1	1	1	1	1	1	1
Greater Kestrel *Falco rupicoloides* 78,88-9,104,109			A	1	1			3		2	1		1	1	1
Lesser Kestrel *Falco naumanni* 64			S	1	1	3		2	1	1	3		2	1	1
Pygmy Falcon *Polihierax semitorquatus* 92-3,106-7,109-11			A	1							2		1	1	
Chukar Partridge *Alectoris chukar* INTRODUCED 31			A	1		2									
Grey-wing Francolin *Franc. africanus* 24,41,44,47,49,63-4,69,75,88	E	E	A	1	1	2		1	1	1	3	2			2
Red-wing Francolin *Francolinus levaillantii* 71			A	1				3							
Orange River Francolin *Francolinus levaillantoides* 115	N		A	1		1							1		
Red-billed Francolin *Francolinus adspersus* 112	N		A	1		1							2		
Cape Francolin (Cape Spurfowl) *Francolinus capensis* 6,15,18,28,47,49	E	N	A	1	1	1		1	1	2	2	2	3		1
Natal Francolin (Natal Spurfowl) *Francolinus natalensis*	N		A	1		1							3		
Red-necked Francolin (Red-necked Spurfowl) *Francolinus afer* 71,119			A	1		1			2						1
Swainson's Francolin (Swainson's Spurfowl) *Francolinus swainsonii*	N		A	1									3		
Common Quail *Coturnix coturnix* 23,46			S	1	1	3		1	1	1	3	3	3	2	1	2
Harlequin Quail *Coturnix delegorguei*			S	1	1			3	3				3		
Helmeted Guineafowl *Numida meleagris*			A	1	1			1	1	1	1	2	2	2	1	1
Common Peafowl *Pavo cristatus* INTRODUCED 31			A	1		2									
Kurrichane Buttonquail (Common Buttonquail) *Turnix sylvatica* 110			A	1									2	2	
Hottentot/Black-rumped Buttonquail *Turn.hottentotta* 6,13-4,22-3,65	E	E	A	1	1			3	2						3
Wattled Crane *Grus carunculatus*				Locally extinct											
Blue Crane *Anthropoides paradiseus* 45-6,58,63-4,72,89	E	N	A	1	1	3		1	1	3	2			2	2	2
Crowned Crane (Grey Crowned Crane) *Balearica regulorum*			A	1					3						
African Rail *Rallus caerulescens* 49,56,69,120			A	1	1	3		1	2	3	3	3		2	1
Corncrake *Crex crex*			S	1	1	3							3		
African Crake *Crex egregia*			A	1	1			3					3		3
Black Crake *Amaurornis flavirostris* 50,84			A	1	1	2		2	2	1	3		3	2	2
Baillon's Crake *Porzana pusilla* 84,120			A	1	1			3	3	3			3	3	1
Striped Crake *Aenigmatolimnas marginalis*			S	1									3		
Red-chested Flufftail *Sarothrura rufa* 56,120			A	1	1	3		2	2	3			3	2	2
Buff-spotted Flufftail *Sarothrura elegans* 19,20,119,122			A	1	1			3	3					1	2
Striped Flufftail *Sarothrura affinis* 19,61,65			A	1	2			2							3
White-winged Flufftail *Sarothrura ayresi*			A	1									3		
Purple Gallinule (Purple Swamphen) *Porphyrio porphyrio* 28-9,42,56,84			A	1	1	1		1	2	1	1	2	2	3	2	1
Lesser Gallinule (Allen's Gallinule) *Porphyrula alleni*			A	1	1	3		3					3		
American Purple Gallinule *Porphyrio martinicus* 50-1,84			W	1		3		3	3	3					
Moorhen (Common Moorhen) *Gallinula chloropus* 42,50			A	1	1	1		1	1	1	2	2	2	1	1
Lesser Moorhen *Gallinula angulata*			S	1				3					3		
Red-knobbed Coot *Fulica cristata* 42			A	1	1	1		1	2	2	2	2	1	1
African Finfoot *Podica senegalensis* 122			A	1										2	
Kori Bustard *Ardeotis kori* 89,106,116			A	1						2	3	1	2	2
Stanley's Bustard *Neotis denhami* 58,63-5,69,72			A	1	1				1					2	1
Ludwig's Bustard *Neotis ludwigii* 7,45,86,89,97,99,104-5,112,125	N		A	1	1	3			3	3	1	1	2	2	1
Blue Korhaan *Eupodotis caerulescens*	E	E	A	1									3		
Karoo Korhaan *Eupodotis vigorsii* 7,63-4,74,78,86,91,94-5,125	E		A	1	1			1	1	1	2	2	2	1
Red-crested Korhaan *Eupodotis ruficrista* 113	N		A	1									1	1	
Southern Black Korhaan *Eu.afra* 12,41,44,49,54,57,64-5,69,77,88,104	E	E	A	1	1	3		1	1	1	3	1		1	1
Northern Black Korhaan/White-quilled *Eu.afraoides* 12,95,109,113-4	E		A	1						1				1	3

Species	sA	SA	Y	WC	NC	1	2	3	4	5	6	7	8A	8B	9
African Jacana *Actophilornis africanus*			A	1	1	3		3	3	3			3	3	3	3
Painted Snipe (Greater Painted Snipe) *Rostratula benghalensis* 29,42			A	1	1	3		3	3				3	3	3	3
European Oystercatcher(Eurasian) *Haematopus ostralegus* 51			A	1		3		3	3							3
African Black Oystercatcher *Hae.moquini* 8,11,22,29,31,43,52,67,101 E			A	1	1	1		1	1				1			1
Ringed Plover (Common Ringed Plover) *Chara. hiaticula* 22,48,53,67			S	1	1	1		1	1	3	3	1	1	1		1
White-fronted Plover *Charadrius marginatus* 22,24,48-9,67,101			A	1	1	1		1	1							1
Chestnut-banded Plover *Charadrius pallidus* 41,46,48-9,52-3,115			A	1		1		1	2		2	3	2	2	3	2
Kittlitz's Plover *Charadrius pecuarius* 49,53			A	1	1	1		1	1		1	2	1	2	2	2
Three-banded Plover *Charadrius tricollaris* 42,44,50,81			A	1	1	1		1	1	1	1	1	1	1	1	1
Mongolian Plover (Lesser Sand-Plover) *Charadrius mongolus* 67			S	1				3	3							3
Greater Sand Plover *Charadrius leschenaultii* 48,67			S	1		3		2	3							3
Caspian Plover *Charadrius asiaticus* 49,51			S	1	1			3							3	
American Golden Plover *Pluvialis dominica* 51			S	1	1	3		3								
Pacific Golden Plover *Pluvialis fulva* 51			S	1	1			3							3	
Grey Plover *Pluvialis squatarola* 47,67,101			S	1	1	1		1	1		3	1	3			2
Crowned Plover (Crowned Lapwing) *Vanellus coronatus*			A	1	1	1		1	1	2	2	2	1	1	1	1
Black-winged Plover (Black-winged Lapwing) *Van.melanopterus* 122			A	1												2
Blacksmith Plover (Blacksmith Lapwing) *Vanellus armatus* 26,49			A	1	1	1		1	1	1	1	1	1	1	1	1
Turnstone (Ruddy Turnstone) *Arenaria interpres* 22,24,47,67,101			S	1	1	1		1	1			1		3		1
Terek Sandpiper *Tringa cinereus* 48,67			S	1		3		2	2				3			3
Common Sandpiper *Tringa hypoleucos* 22,67,84			S	1	1	1		2	1	2	2	2	3	2	2	2
Green Sandpiper *Tringa ochropus* 51			S	1	1			3					3			
Wood Sandpiper *Tringa glareola* 28,42			S	1	1	1		2	2	2	2	2	2	2	2	1
Redshank (Common Redshank) *Tringa totanus* 48,51			S	1	1	3		3	3			3		3	3	
Marsh Sandpiper *Tringa stagnatilis* 47			S	1	1	2		1	2	3	3	2	3	2	1	
Greenshank (Common Greenshank) *Tringa nebularia* 47			S	1	1	1		1	1	1	2	1	2	1	1	
Lesser Yellowlegs *Tringa flavipes* 50-1			S	1				3								
Greater Yellowlegs *Tringa melanoleuca* 51			S	1		3										
Knot (Red Knot) *Calidris canutus* 47			S	1	1	1		1	3			2				3
Curlew Sandpiper *Calidris ferruginea* 47-8,53,67			S	1	1	1		1	1	2	2	1	3			2
Dunlin *Calidris alpina* 51			S	1		3										
Little Stint *Calidris minuta* 29,42,47,49,53			S	1	1	1		1	2	2	2	2	2	1	2	
Red-necked Stint *Calidris ruficollis* 51			S	1		3										
White-rumped Sandpiper *Calidris fuscicollis* 51			S	1		3	3									
Baird's Sandpiper *Calidris bairdii* 51			S	1		3										
Pectoral Sandpiper *Calidris melanotos* 51			S	1	1	3		3							3	3
Sanderling *Calidris alba* 24,47,67,101			S	1	1	1		1	1		3	1	3			2
Buff-breasted Sandpiper *Tryngites subruficollis* 51			S	1		3										
Broad-billed Sandpiper *Limicola falcinellus* 51			S	1		3										
Ruff *Philomachus pugnax* 42,53			S	1	1	1		1	2	3	3	2	2	1	2	
Great Snipe *Gallinago media*			S	1		3										
Ethiopian Snipe (African Snipe) *Gallinago nigripennis* 29,42,46,84			A	1	1	2		1	2						2	3
Black-tailed Godwit *Limosa limosa* 51			S	1	1	3		3				3		3	3	
Bar-tailed Godwit *Limosa lapponica* 48,50,54,67			S	1	1	3		1	2			2			2	
Hudsonian Godwit *Limosa haemastica* 51			S	1		3										
Curlew (Eurasian Curlew) *Numenius arquata* 54			S	1	1	3		1	2			2				
Whimbrel *Numenius phaeopus* 22,24,47			S	1	1	1		1	1			1	3		1	
Grey Phalarope *Phalaropus fulicaria* 36,51			S	1		3	3	3		3						
Red-necked Phalarope *Phalaropus lobatus* 51,54			S	1	1	3		3								
Wilson's Phalarope *Phalaropus tricolor* 51			S	1		3		3								
Avocet (Pied Avocet) *Recurvirostra avosetta* 28,44,54,89			A	1	1	1		1	2	2	1	1	2	1	3	
Black-winged Stilt *Himantopus himantopus* 42,49,54			A	1	1	1		1	1	1	1	1	1	1	1	1
Crab Plover *Dromas ardeola* 51			A	1				3								3
Spotted Dikkop (Spotted Thick-Knee) *Burhinus capensis* 46			A	1	1	1		1	1	1	1	1	1	1	1	2
Water Dikkop (Water Thick-Knee) *Burhinus vermiculatus* 83-4			A	1	1	1	2	2	2	1	1	3			2
Burchell's Courser *Cursorius rufus* 86,90-1,93,96,109,112	N		A	1	1			3	3	3	1	3	2	2	2
Temminck's Courser *Cursorius temminckii* 110			A	1	1	3								2	2	2
Double-banded Courser *Smutsornis africanus* 78,91,106,112,115,125			A	1	1			3	2		1	3	1	1	2	
Bronze-winged Courser *Rhinoptilus chalcopterus*			A	1										3	3	
Red-winged Pratincole (Collared Pratincole) *Glareola pratincola*			A	1		3		3	3							
Black-winged Pratincole *Glareola nordmanni* 115			S	1	1										3	3	3
Greater Sheathbill *Chionis alba* 50-1			W	1		3		3								
Arctic Skua *Stercorarius parasiticus* 31,35-6,39,54			S	1	1	1	1	1	2							2
Long-tailed Skua *Stercorarius longicaudus* 35-6,39			S	1	1	3	2									
Pomarine Skua *Stercorarius pomarinus* 31,35-6,39			S	1	1	2	2	3	3							
Subantarctic Skua *Catharacta antarctica* 31,35-6,39			A	1	1	1	1	1	2							3
South Polar Skua *Catharacta maccormicki* 37			W	1		3										
Kelp Gull *Larus dominicanus* 13			A	1	1	1	1	1	1	2		1			1
Lesser Black-backed Gull *Larus fuscus*			A	1											3	
Grey-headed Gull *Larus cirrocephalus* 42,54			A	1	1	1		1	2	1	2	2	3	1	3	
Hartlaub's Gull *Larus hartlaubii* 8,31 E			A	1	1	1	1	1	1			1			3
Franklin's Gull *Larus pipixcan*			S	1		3		3								
Sabine's Gull *Xema sabini* 31,35-6,39			S	1	1	1	1	3								
Black-headed Gull *Larus ridibundus*			S	1		3										
Black-legged Kittiwake *Rissa tridactyla* 37			S	1		3										
Caspian Tern *Sterna caspia* 42,53-4,56,67,101			A	1	1	1		1	1	3						1
Swift Tern *Sterna bergii* 22,24,29,31,39,55,67			A	1	1	1	1	1	1			1			1
Lesser Crested Tern *Sterna bengalensis*			S	1		3			3							
Sandwich Tern *Sterna sandvicensis* 22,24,29,67			S	1		1	1	1	1			1			1

	sA	SA	Y	WC	NC	1	2	3	4	5	6	7	8A	8B	9
Common Tern *Sterna hirundo* 22,24,31,67			S	1	1	1	1	1	1			1			1
Arctic Tern *Sterna paradisaea* 35-6			S	1	1	2	1	2	2			2			3
Antarctic Tern *Sterna vittata* 14,21-2,36			W	1		1	2	2	3						
Roseate Tern *Sterna dougallii*			A	1		3			3						
Sooty Tern *Sterna fuscata*			S	1				3							
Bridled Tern *Sterna anaethetus*			S	1				3							
Damara Tern *Sterna balaenarum* 8,58-9,67-8,97,101	N		S	1	1	3		3	1			1			
Little Tern *Sterna albifrons* 50,54			S	1	1	3		1	3			3			3
Black Tern *Chlidonias niger*			S	1	1	3		3	3			3			
Whiskered Tern *Chlidonias hybridus*			A	1	1	2		2	2	2	3	3	3	2	3
White-winged Tern *Chlidonias leucopterus* 28,42,84			S	1	1	1		1	1	1	3	3	3	1	1
Common Noddy *Anous stolidus*			A	1				3							
Namaqua Sandgrouse *Pterocles namaqua* 65,77-8,91,95,104,112,125	N		A	1	1	2		2	2	1	1	1	1	1	1
Burchell's Sandgrouse *Pterocles burchelli* 106-7,110,112,116	N		A	1								1	1	1	
Yellow-throated Sandgrouse *Pterocles gutturalis*			A	1									3		
Double-banded Sandgrouse *Pterocles bicinctus* 112	N		A	1							3	3	2		
Feral Pigeon *Columba livia* INTRODUCED 31			A	1	1	1		1	1	1	1	1	1	1	1
Rock Pigeon (Speckled Pigeon) *Columba guinea* 20			A	1	1	1		1	1	1	1	1	1	1	1
Rameron Pigeon (African Olive-Pigeon) *Columba arquatrix* 17,19			A	1	1	1		3	2						1
Red-eyed Dove *Streptopelia semitorquata* 26			A	1	1	1		1	1	2	2	2	1		1
Cape Turtle Dove (Ring-necked Dove) *Streptopelia capicola*			A	1	1	1		1	1	1	1	1	1	1	1
European Turtle Dove *Streptopelia turtur*			A	1									3		3
Laughing Dove *Streptopelia senegalensis*			A	1	1	1		1	1	1	1	1	1	1	1
Namaqua Dove *Oena capensis* 45,50			A	1	1	1		2	1	1	1	1	1		1
Green-spotted Dove (Emerald-spotted Dove) *Turtur chalcospilos*			A	1											3
Tambourine Dove *Turtur tympanistria* 68,119			A	1					1	3					1
Cinnamon Dove (Lemon Dove) *Columba larvata* 17,19,119			A	1	1	1									1
Rose-ringed Parakeet *Psittacula krameri* INTRODUCED			A	1		3									
Rosy-faced Lovebird *Agapornis roseicollis* 93-4,111	N		A	1							1	3	2		
Knysna Lourie (Knysna Turaco) *Tauraco corythaix* 7,12,117-119,125	E	E	A	1		3									1
Grey Lourie (Grey Go-away Bird) *Corythaixoides concolor*			A	1	1				3					3	
European Cuckoo (Common Cuckoo) *Cuculus canorus*			S	1	1				3				3		3
African Cuckoo *Cuculus gularis*			S	1									2	1	
Red-chested Cuckoo *Cuculus solitarius* 18-9			S	1	1	1		3	1	2					3	1
Black Cuckoo *Cuculus clamosus*			S	1	1			3					3	2	3
Great Spotted Cuckoo *Clamator glandarius* 110			S	1	1	3		3	3				2	3	3
Striped Cuckoo (Le Vaillant's Cuckoo) *Oxylophus levaillantii*			S	1					1					3	
Jacobin Cuckoo *Oxylophus jacobinus*			S	1	1	3		3		3			2	1	3
Emerald Cuckoo (African Emerald Cuckoo) *Chryso. cupreus* 8,117,122			S	1											1
Klaas's Cuckoo *Chrysococcyx klaas* 45,69			A	1	1	1		1	1	2				3	1
Diederik Cuckoo *Chrysococcyx caprius* 45			S	1	1	2		2	2	3	3	3	2	1	1
Burchell's Coucal *Centropus burchellii* 20			A	1	1	1		2	2	3	3			2	1
Barn Owl *Tyto alba* 109			A	1	1	2		1	1	2	2	2	1	1	3
Grass Owl (African Grass Owl) *Tyto capensis*			A	1	1									3	3
Wood Owl (African Wood Owl) *Strix woodfordii* 19,20,68,118,122			A	1	1			1							1
Marsh Owl *Asio capensis*			A	1	1	3		2	3	3		3		2	3
Scops Owl (African-Scops Owl) *Otus senegalensis* 109			A	1									2	3	
White-faced Owl (White-faced Scops-Owl) *Otus leucotis* 109,111			A	1									1	2	
Pearl-spotted Owl (Pearl-spotted Owlet) *Glaucidium perlatum* 109,111			A	1									1	2	
Cape Eagle-Owl *Bubo capensis* 13,56,63,81,88,94-5,97,99,103-4,123			A	1	1	3		3	2	3	2	1			3
Spotted Eagle-Owl *Bubo africanus* 16,20,54,109			A	1	1	1		1	1	1	1	1	1	1	1
Giant Eagle-Owl (Verreaux's Eagle-Owl) *Bubo lacteus* 106-7,111,116			A	1	1			3				1	1	2	3
European Nightjar *Caprimulgus europaeus*			S	1	1			3		3			3	2	3
Fiery-necked Nightjar *Caprimulgus pectoralis* 20,69			A	1	1	1		1	1	2	3		3	3	1
Rufous-cheeked Nightjar *Caprimulgus rufigena* 95			S	1	1				3	2	2	3	1	1	1
Freckled Nightjar *Caprimulgus tristigma* 103			A	1	1	3		3	3	2	3	1	1		1
European Swift (Eurasian Swift) *Apus apus* 110			S	1	1			3		3	1	2	1		2
African Black Swift *Apus barbatus* 21,23,52,62,84			A	1	1	1		1	1	1	1	3	3	2	2
Bradfield's Swift *Apus bradfieldi* 94,101,106,113	N		A	1							2	1	2	1	
Pallid Swift *Apus pallidus*			S	1										3	
White-rumped Swift *Apus caffer*			S	1	1	2		1	1	2	2	2	2	1	1
Horus Swift *Apus horus* 64,66			A	1	1	3		3	1	3				3	3
Little Swift *Apus affinis* 46			A	1	1	2		1	1	1	1	1	1	1	1
Alpine Swift *Tachymarptis melba* 21,62			A	1	1	1		1	1	1	1	2	1	2	1
Palm Swift (African Palm Swift) *Cypsiurus parvus* 94			A	1							1	3	3	2	
Speckled Mousebird *Colius striatus* 18,24			A	1	1	1		1	1	2					1
White-backed Mousebird *Colius colius* 25,44,47,54,74,80,89,98	E		A	1	1	2		1	2	1	1	1	1	1	1
Red-faced Mousebird *Urocolius indicus* 54			A	1	1	2		1	1	2	1	1	1	1	1
Narina Trogon *Apaloderma narina* 8,58,69,71,117,119,125			A	1					1						1
Pied Kingfisher *Ceryle rudis* 42,67			A	1	1	1		1	1	1	1	1	1	1	1
Giant Kingfisher *Megaceryle maxima* 69,113			A	1	1	2		2	1	3	2	2	1		1
Half-collared Kingfisher *Alcedo semitorquata* 117,119,122			A	1		3		3	3	3					1
Malachite Kingfisher *Alcedo cristata* 56,83,84,120			A	1	1	1		1	1	2	2			1	1	1
Mangrove Kingfisher *Halcyon senegaloides*			A	1					3						
Brown-hooded Kingfisher *Halcyon albiventris* 68			A	1	1	3		3	1					3	1
Grey-hooded Kingfisher (Grey-headed Kingfisher) *Halcyon leucocephala*			A	1	1	3		3					3	3	
Striped Kingfisher *Halcyon chelicuti*			A	1									3	3	
European Bee-eater *Merops apiaster* 46,90,104			S	1	1	3		1	3	2	1	1	3		1
Blue-cheeked Bee-eater *Merops persicus*			S	1	1	3								2	3

Species	sA	SA	Y	WC	NC	1	2	3	4	5	6	7	8A	8B	9
Carmine Bee-eater *Merops nubicoides*				1				3							
White-fronted Bee-eater *Merops bullockoides* 113			A	1	1	3					3	3	3	2	
Little Bee-eater *Merops pusillus*			A	1				3							
Swallow-tailed Bee-eater *Merops hirundineus* 94,109,111			A		1						1	1	1	1	
White-throated Bee-eater *Merops albicollis*			A	1	1	3							3		3
European Roller *Coracias garrulus*			S	1	1	3		3	3				3	2	3
Lilac-breasted Roller *Coracias caudata* 109-10			A	1									1	1	
Purple Roller (Rufous-crowned Roller) *Coracias naevia*			A	1									2	3	
African Hoopoe *Upupa africana* 45,114			A	1	1	2		1	3	3	2	1	1	1	1
Red-billed Woodhoopoe (Green Woodhoopoe) *Phoeniculus purpureus*			A	1	1								3	3	1
Scimitar-billed W.hp. (Common Scimitarbill) *Rhinopomastus cyanomelas*			A	1	1	3					2	3	1		1
Grey Hornbill (African Grey Hornbill) *Tockus nasutus*			A	1									2	1	
Yellow-billed Hornbill (Southern Y-b H.) *Tockus leucomelas* 110,112 N			A	1									1	1	
Crowned Hornbill *Tockus alboterminatus*			A		1										3
Black-collared Barbet *Lybius torquatus*			A	1										3	
Acacia Pied Barbet *Tricholaema leucomelas* 26,45,52,67,80,93,112	N		A	1	1	1		1	1	1	1	1	1	1	1
Red-fronted Tinker Barbet (Red-fronted Tinkerbird) *Pogoniulus pusillus*			A	1											3
Crested Barbet *Trachyphonus vaillantii* 113			A	1	1			3				3	2	1	
Greater Honeyguide *Indicator indicator* 45,65			A	1	1	3		2	2	3				2	2
Scaly-throated Honeyguide *Indicator variegatus* 119,122			A	1											1
Lesser Honeyguide *Indicator minor* 26,65,69,114			A	1	1	3		1				1	2	2	2
Sharp-billed Honeygde (Brown-backed H.bird) *Prodotiscus regulus* 65			A	1		3			3	3					3
Ground Woodpecker *Geocolaptes olivaceus* 19,21,61-3,84,99,123-4 E		E	A	1	1	1	1		1	1	1	3	1		1
Bennett's Woodpecker *Campethera bennettii*			A	1										3	
Golden-tailed Woodpecker *Campethera abingoni* 109-10,112			A	1							3		1		1
Knysna Wdpckr *Campethera notata* 7,58,64-6,69,71-2,117,120,122 E		E	A	1					1	3					1
Cardinal Woodpecker *Dendropicos fuscescens* 45,49,69,123			A	1	1	3		1	1	3	3	3	1		1
Bearded Woodpecker *Thripias namaquus*			A	1									3	3	
Olive Woodpecker *Mesopicos griseocephalus* 63,68,71,81,119			A	1		3			1	2					1
Melodious Lark *Mirafra cheniana*	E		A	1											
Monotonous Lark *Mirafra passerina* 106,110,112	N		A	1									2	2	
Rufous-naped Lark *Mirafra africana* 113			A	1											1
Clapper Lark *Mirafra apiata* 7,13,24,45,55-6,58,63-4,69,88,104,115-6	N		A	1	1	2		1	1		2	1	1	1	1
Fawn-coloured Lark *Mirafra africanoides* 94-5,110-1			A	1							2		1	1	
Sabota Lark *Mirafra sabota* 13,93-4,114	N		A	1							1	1	1	1	
Cape Long-billed Lark *Certhil. curvirostris* 13,41,52,54-6-97,101,104	E	E	A	1	1			1				1			
Agulhas Long-billed Lark *Certhilauda brevirostris* 6,12-3,58-9,63-4,73 E		E	A	1	1				1						2
Karoo Long-billed Lark *Certhilauda subcoronata* 7,13,89,93-4,102,112 E			A	1	1					3	1	2	1	2	1
Short-clawed Lark *Certhilauda chuana* 8,106,113	E		A	1										2	
Karoo Lark *Mirafra albescens* 7,44,54-5,74,89,90,97,99,100,103,125	E	E	A	1	1				1	1	1	1			1
Barlow's Lark *Mirafra barlowi* 7,12,97,99,100-2	E		A	1							1	1			
Red Lark *Mirafra burra* 7,12,86,89,90-1,95-6,102	E	E	A	1							1	1			
Dusky Lark *Pinarocorys nigricans*			S										3	3	
Spike-heeled Lark *Chersomanes albofasciata* 78,91,104,111-3,115	N		A	1	1			1	1	1	1	1	1	1	1
Red-capped Lark *Calandrella cinerea* 45-6,52,64,78,89,115			A	1	1	1		1	1	1	1	1	1	1	1
Pink-billed Lark *Spizocorys conirostris* 92,106-7,109,111,115			A	1							3		1	2	
Sclater's Lark *Spizocorys sclateri* 7,86,90-2,94-6	E		A	1	1	3						3			
Stark's Lark *Eremalauda starki* 86,92-3,95,112	N		A	1								1	3	1	3
Thick-billed Lark (Large-billed Lark) *Galerida magnirostris* 46,54-5,78 E			A	1	1			1				1	1	1	2
Chestnut-backed Finchlark (Chestnut-bkd Sparrow-Lark) *Erem. leucotis*			A								3		3	3	1
Grey-backed Finchlark (G.-b. Spa. La.) *Eremopterix verticalis* 52,95	N		A	1	1	2		2	3		1	1	1	1	1
Black-eared Finchlark (Black-eared Spa. La.) *Erem. australis* 7,86,96 E			A	1	1				3		1	3	1	2	2
European Swallow (Barn Swallow) *Hirundo rustica* 45,47			S	1	1	1			1	1	1	1	1	1	1
White-throated Swallow *Hirundo albigularis* 28,43,45,47,66,84			S	1	1	1		1	1	1	1	1	1	1	1
Pearl-breasted Swallow *Hirundo dimidiata* 45,47,55,66,69			S	1	1	3		1		2	3	3	3	1	1
Red-breasted Swallow *Hirundo semirufa* 113			S	1											1
Greater Striped Swallow *Hirundo cucullata* 45	N		S	1	1	1		1	1	1	1	1	1	1	1
Lesser Striped Swallow *Hirundo abyssinica*			S	1		3									3
South African Cliff Swallow *Hirundo spilodera* 89,115	N		A	1	1			3			1		3	1	
Rock Martin *Hirundo fuligula* 21,62,80			A	1	1				1	1	1	1	1	1	1
House Martin (Common House Martin) *Delichon urbica*			S	1	1	3		3	3		3	3	3	3	2
Sand Martin *Riparia riparia*			S	1	1	3		3	3			3			
Brown-throated Martin (Plain Martin) *Riparia paludicola* 28,43,50,66			S	1	1	1		1	1	1	1	1	1	1	1
Banded Martin *Riparia cincta* 45			S	1	1	3		1	3	3					3
Black Sawwing Swallow (Blck Sw-wng) *Psalidoprocne holomelas* 18,62			S	1	1	1		3	1						1
Black Cuckooshrike *Campephaga flava* 66			A	1				3	3						1
Grey Cuckooshrike *Coracina caesia* 71,119			A	1											1
Fork-tailed Drongo *Dicrurus adsimilis*			A	1	1				1				1	1	1
European Golden Oriole (Eurasian Golden Oriole) *Oriolus oriolus*			S	1	1			3	3	3			3	3	2	3
Black-headed Oriole (Eastern Blk-headed Or.) *Oriolus larvatus* 119			A	1	1			3							1
Black Crow (Cape Crow) *Corvus capensis* 64,88,99,104			A	1	1	1		3	1	3	1	1	1	2	1
Pied Crow *Corvus albus* 43			A	1	1	1					1	1	1	1	1
House Crow *Corvus splendens* INTRODUCED 31,60			A	1				3							
White-necked Raven *Corvus albicollis* 21,81			A	1	1	1		2	1	3					1
Southern Grey Tit *Parus afer* 41,52,54,56,74,77,80,97,99,101-2,104	E	N	A	1	1			1		1	1	1	1		
Ashy Tit *Parus cinerascens* 8,109,111-2,114	N		A	1							3		1	1	1
Cape Penduline Tit *Anthoscopus minutus* 41,44,52,54,78,81,89,101 N			A	1	1	3					1	1	1	1	1
Pied Babbler (Southern Pied Babbler) *Turdoides bicolor* 112	N		A	1									3	3	
Cape Bulbul *Pycnonotus capensis* 6,24,28,44,47,62,66,69,83,123	E	E	A	1	1	1		1	1	1		1			1

Species	sA	SA	Y	WC	NC	1	2	3	4	5	6	7	8A	8B	9
Red-eyed Bulbul (African Red-eyed Bulbul) *Pycnonotus nigricans* 123	N		A	1	1						1	1	1	1	1
Terrestrial Bulbul (Terrestrial Brownbul) *Phyllastrephus terrestris* 71,118			A	1											1
Sombre Bulbul (Som. Greenbul) *Andropadus importunus* 17,19,71,119			A	1					1	3					1
Olive Thrush *Turdus olivaceus* 13,17,111			A	1	1	1		2	1	1	2	2	1		1
Groundscraper Thrush *Psophocichla litsitsirupa*			A		1								3	1	
Cape Rock-Thrush *Monticola rupestris* 24,61-2,81,123	E	E	A	1	1	1		3	1	1					1
Sentinel Rock-Thrush *Monticola explorator* 24,61,123	E	E	A	1	1	3			2	3					1
Short-toed Rock-Thrush *Monticola brevipes* 13,94,112,117,123-4	N		A	1	1						2	3	2	1	1
Mountain Chat (Mountain Wheatear) *Oenanthe monticola* 76,89,102	N		A	1	1	3		1	3	1	1	1	1	1	1
Capped Wheatear *Oenanthe pileata* 45-6,54,64-5			A	1	1	1		1	1	1	1	1	1		1
Familiar Chat *Cercomela familiaris* 21,24,45,61,89			A	1	1	1		1	1	1	1	1	1		1
Tractrac Chat *Cercomela tractrac* 7,74,78,89,101	N		A	1	1					1	1	1	1		1
Sickle-winged Chat *Cercomela sinuata* 41,52,77-8,89,124	E	N	A	1	1	3		1	3	2	1	2	2	2	1
Karoo Chat *Cercomela schlegelii* 7,77-8,89,93,103	N		A	1	1	3		3	1	1	1	1			1
Mocking Chat (Mocking Cliff-Chat) *Thamnolaea cinnamomeiventris*			A	1											3
Anteating Chat (Southern A. Ch.) *Myrmecocichla formicivora* 52,111	E		A	1	1	1		1	1	1	1	1	1		1
Stonechat (Common Stonechat) *Saxicola torquata* 49,54			A	1	1	3		1	1	1	2	1	3	1	1
Chorister Robin *Cossypha dichroa* 7,117-8,125	E	E	A	1											1
Cape Robin (Cape Robinchat) *Cossypha caffra* 18,23,44,47			A	1	1	1		1	1	1	1	1	1		1
Starred Robin (White-starred Robin) *Pogonocichla stellata* 120,122			A	1											1
Cape Rockjumper *Chaetops frenatus* 6,58,60-2,65,73,81,117,123	E	E	A	1		3		1							1
Karoo Robin (Karoo Scrubrobin) *Cercotrichas coryphaeus* 44,47,89	E		A	1	1	2		1	1	1	1	1	1	1	1
Kalahari Robin (Kalahari Scrubrobin) *Cercotrichas paena* 8,109,114	N		A	1								1	1	1	1
Garden Warbler *Sylvia borin*			S	1	1			3						3	3
Whitethroat (Greater Whitethroat) *Sylvia communis*			S	1										3	
Titbabbler (Chestnut-vented Titb.) *Parisoma subcaeruleum* 44,47,76	N		A	1	1	3		1	2	1	1	1	1	1	1
Layard's Titbabbler *Parisoma layardi* 44,54,79,80,88-9,97,102,122-4	E		A	1	1	1		1	1	1	1	1	2	1	1
Icterine Warbler *Hippolais icterina*			S	1		3							3	2	
Great Reed-Warbler *Acrocephalus arundinaceus* 119			S	1	1								3	3	3
African Marsh Warbler (Afr. Reed-Warbler) *Acro. baeticatus* 28,84,89			S	1	1	2		2	2	1	1	1	1	1	1
European Marsh Warbler (Marsh Warbler) *Acrocephalus palustris*			S	1	1										3
European Sedge Warbler (Sedge War.) *Acrocephalus schoenobaenus*			S	1	1			3				3	3	3	
Cape Reed Warbler (Lesser Swamp-War.) *Acro. gracilirostris* 28,42			A	1	1	1		1	1	1	1	1	1	1	1
African Sedge War. (Little Rush War.) *Bradypterus baboecala* 28,49			A	1	1	1		1	1				3	3	1
Knysna Warbler *Bradypterus sylvaticus* 7,11,14,17-9,20,32,69,71,120-1	E	E	A	1		1		2							1
Victorin's Warbler *Bradypterus victorini* 6,58,63,69,71,73,81,122	E	E	A	1					1	1					1
Willow Warbler *Phylloscopus trochilus*			S	1	1	3		3	2	3	3	1	3	1	1
Yellow-throated Warbler (Y-thr. Woodland-War.) *Phyllosc. ruficapilla* 71			A	1											1
Bar-throated Apalis *Apalis thoracica* 20,44,47,54,62,66,69,71,81-3,118			A	1	1	2		1	1	1	1	1	1		1
Long-billed Crombec *Sylvietta rufescens* 44-5,47,54,56			A	1	1	3		1	1	1		1	2	1	1
Yellow-bellied Eremomela *Eremomela icteropygialis* 52,55,78,81,90			A	1	1	2		3	2	2	1	1	1	1	1
Karoo Eremomela *Eremomela gregalis* 7,74,77-8,85-6,90,98-9,102	E		A	1					1	1		1			1
Green-backed Bleating War.(Gr.-b. Camaroptera) *Cama. brachyura* 118			A	1											1
Barred Warbler (Barred Wren-Warbler) *Calamonastes fasciolatus*	N		A		1									3	
Cinnamon-breasted Warbl. *Euryptila subcinnamomea* 11,79,85,89	E		A	1		3			1	1		1			1
Grassbird *Sphenoeacus afer* 18,20,24,44,47,61,63,123	E	N	A	1	1	1		1	1	1	1	1	1		1
Fan-tailed Cisticola *Cisticola juncidis* 94,101			A	1	1	2		2	1				1	1	1
Desert Cisticola *Cisticola aridulus* 110			A	1	1	3							1	1	3
Cloud Cisticola *Cisticola textrix* 13,24,41,46,57,64	N	N	A	1	1	1		1	3				3	3	3
Grey-backed Cisticola *Cisticola subruficapillus* 23,44,47,49,77	N		A	1	1	1		1	1	1	1	1	2	2	1
Wailing Cisticola *Cisticola lais*			A	1											3
Tinkling Cisticola *Cisticola rufilatus*			A		1								3	3	
Le Vaillant's Cisticola *Cisticola tinniens* 18,28,43,49,50,54,81			A	1	1	1		1	1	1	2	3	2		1
Neddicky *Cisticola fulvicapillus* 20,61,63,123			A	1	1	3		1	1			1	1	1	1
Black-chested Prinia *Prinia flavicans* 89,112	N		A	1	1						1	2	1	1	1
Karoo (old: Spotted) Prinia *Prinia maculosa*	E	N	A	1	1	1		1	1	1	1	1	1		1
Namaqua Warbler *Phragmacia substriata* 7,52,74,76,78,85,89,110,123	E	N	A	1					1	1	1	1	1	2	1
Rufous-eared Warbler *Malcorus pectoralis* 7,55,77,89,104	E		A	1	1	1		1	1	1	1	1	1	1	1
Spotted Flycatcher *Muscicapa striata*			S	1	1	3		3	3	3	3	1	1		2
Dusky Flycatcher (African Dsky Flyc.) *Muscicapa adusta* 17,19,62,71,119			A	1		1		3	1	1					1
Marico Flycatcher *Bradornis mariquensis* 8,109	N		A	1									1		1
Chat Flycatcher *Bradornis infuscatus* 55,89,104,111-2,124	N		A	1		2			2	1	1	1	2	1	1
Fiscal Flycatcher *Sigelus silens* 24,45,67,69,81,83,114,123	E	N	A	1	1	1		1	1	3	2	1	1		1
Cape Batis *Batis capensis* 7,17,19,62,71,83,119	E		A	1		3		1	1						1
Pririt Batis *Batis pririt* 74,77-9,80,85,88,90,92-3,103-4,112-3,123	N		A	1	1	1			1	1	1	1	1	1	1
Fairy Flycatcher *Stenostira scita* 56,74,76,79,80,88-9,92,97,102-3,122-3	E	N	A	1		1		1	1	1	1	1	1	1	1
Blue-mantled Fly. (B-m. Crested-Fly) *Trochocercus cyanomelas* 63,71,119			A	1		3									1
Paradise Flycatcher (African Parad.-Flyc.) *Terpsiphone viridis* 17,62,71,84			S	1	1	3		1	1	3				3	3
African Pied Wagtail *Motacilla aguimp*			A	1	1	3						1	1	1	3
Cape Wagtail *Motacilla capensis* 49			A	1	1	1		1	1	1	1	1	1	1	1
Yellow Wagtail *Motacilla flava*			S	1	1	3							3	3	3
Grey Wagtail *Motacilla cinerea*			S	1	1	3		3							
Grassveld Pipit (African Pipit) *Anthus cinnamomeus* 45-6,114-5			A	1	1	1		1	1	1	1	1	1	1	1
Long-billed Pipit *Anthus similis* 64,115,124			A	1	1	2		2	2	2		1	2	2	2
Plain-backed Pipit *Anthus leucophrys* 22,24,114-5			A	1	1	2		2	3				1	2	2
Buffy Pipit *Anthus vaalensis* 113-5			A	1	1								2	1	3
Long-tailed Pipit *Anthus longicaudatus* 12,106,113-5	?		W											1	
African Rock Pipit (Yellow-tufted P.) *Anthus crenatus* 112,117,123-125	E	E	A	1					3	3	3			1	
Yellow-breasted Pipit *Hemimacronyx chloris*	E	E	A	1				3							

Species	sA	SA	Y	WC	NC	1	2	3	4	5	6	7	8A	8B	9
Orange-throated Longclaw (Cape L.) *Macronyx capensis* 24,46,64	E		A	1	1	1		1	2					2	1
Lesser Grey Shrike *Lanius minor* 110			S	1	1			3			3		1	1	3
Fiscal Shrike (Common Fiscal) *Lanius collaris* 18,43			S	1	1	1	1	1	1	1	1	1	1	1	1
Red-backed Shrike *Lanius collurio*			S	1	1	3		3		3		3	1	1	1
Southern Boubou *Laniarius ferrugineus* 18,23-5,62,66,69	E	N	A	1	1	1		2	1	1					1
Crimson-breasted Shrike *Laniarius atrococcineus* 8,109,113-4	N		A		1										
Puffback (Black-backed Puffback) *Dryoscopus cubla*			A												
Brubru *Nilaus afer* 111,113			A												
Southern Tchagra *Tchagra tchagra* 58,64-7,69,117,123	E	N	A	1	1	1		1							1
Three-streaked Tchagra (Brown-crowned Tch.) *Tchagra australis*			A		1										
Bokmakierie *Telophorus zeylonus* 44,47,89,102,104	N		A	1	1	1		1	1	1	1	1	1	1	1
Olive Bushshrike *Telophorus olivaceus* 69,71,119			A	1											
White-crowned Shrike (Southern W-cr. Sh.) *Eurocephalus anguitimens*	N		A											3	
European Starling *Sturnus vulgaris* INTRODUCED 31			A	1	1	1	1	2	2					3	1
Indian Myna (Common Myna) *Acridotheres tristis* INTRODUCED			A	1		3									
Pied Starling (African Pied Starling) *Spreo bicolor* 44,47,64	E	E	A	1	1	3		1	1		1	1	1	1	1
Wattled Starling *Creatophora cinerea* 47,50			A	1	1	3		1	1	3	2	2	1	1	1
Plum-coloured Starling (Violet-backed Starl.) *Cinnyricinclus leucogaster*			A		1									3	3
Burchell's Starling *Lamprotornis australis*	N		A											2	3
Glossy Starling (Cape Glossy-Starling) *Lamprotornis nitens* 98,103	N		A	1	1	1								3	1
Black-bellied Starling *Lamprotornis corruscus* 119			A					3							1
Red-winged Starling *Onychognathus morio* 20,23,76			A	1	1	1	1	1		1	1	1	1	1	1
Pale-winged Starling *Onychog. nabouroup* 7,76,80,88,93,102,112,123	N		A	1	1	1	1	1		1	1	1	1	1	1
Cape Sugarbird *Promerops cafer* 6,14-5,17,24,33,61,83,123	E	E	A	1	1	3		1	1	1					1
Malachite Sunbird *Nectarinia famosa* 24,44,47,62,69,77,83,101,123			A	1	1	1		1	1	1	1	1	1		1
Orange-breasted Sunbird *Nectarinia violacea* 6,15,17,20,33,61,83,123	E	E	A	1	1	1		1							1
Marico Sunbird *Nectarinia mariquensis*			A												
Lesser Double-collared Sunbird (Southern D-c Su.) *N. chalybea* 18,20,62	E	N	A	1	1	1		1		1	1	1	3		1
Greater Double-collared Sunbird *Nectarinia afra* 70-1,121	E	N	A	1				1						3
White-bellied Sunbird *Nectarinia talatala*			A		1										
Dusky Sunbird *Nectarinia fusca* 80,90,94,97-9,102,112,123	N		A	1	1	3		2	1	1	1			1	1
Black Sunbird (Amethyst Sunbird) *Nectarinia amethystina* 70			A	1		3		2							1
Collared Sunbird *Anthreptes collaris*			A		1										3
Cape White-eye *Zosterops pallidus* 13,94,101,112,114	E		A	1	1	1	1	1	1	1	1	1	3		1
Red-billed Buffalo-Weaver *Bubalornis niger*			A		1										
White-browed Sparrow-Weaver *Plocepasser mahali* 109			A							2		1	1	1	
Sociable Weaver *Philetairus socius* 92-3,109-10	E		A	1	1	1	1	1	1	1	1	1			1
House Sparrow *Passer domesticus* INTRODUCED 31	N		A	1		1		1	1	1	1	1	2	2	
Great Sparrow *Passer motitensis* 110	N		A	1	1									3	
Cape Sparrow *Passer melanurus* 45,49	N		A	1	1	2		1	1	1	1	1	1	1	1
Grey-headed Sparrow (Southern Grey-hded Sp.) *Passer griseus* 68	N		A	1	1			1		3	3	1	1	2	
Yellow-throated Sparrow (Yellow-thrtd. Petronia) *Petronia superciliaris*			A	1										3	
Scaly-feathered Finch *Sporopipes squamifrons* 95,110	N		A	1	1			1	2	1				3	
Thick-billed Weaver *Amblyospiza albifrons*			A	1										2	
Cape Weaver *Ploceus capensis* 44,47,49	E	E	A	1	1	1		1	1	1	1	1	1	1	1
Masked Weaver (Southern Masked-Weaver) *Ploceus velatus* 45			A	1	1	1	1	1	1	1	1	1	1	1	1
Red-billed Quelea *Quelea quelea* 94,101			A	1	1									1	
Red Bishop (Southern Red Bishop) *Euplectes orix* 84,94			A	1	1	3		1	1	1	1	1	3	2	
Golden Bishop (Yellow-crowned Bishop) *Euplectes afer* 115			A	1	1	1		1	1	1				1	
Yellow-rumped Widow (Yellow Bishop) *Euplectes capensis* 56,61-2			A	1									2	1	
Melba Finch (Green-winged Pytilia) *Pytilia melba* 112			A									3	3	3	2
Red-billed Firefinch *Lagonosticta senegala* 123			A	1	1									3	
Blue Waxbill *Uraeginthus angolensis*			A										1	1	
Violet-eared Waxbill *Uraeginthus granatinus* 110,112	N		A										1	1	
Common Waxbill *Estrilda astrild* 18,20,43			A	1	1	1	1	1	1	1	1	1	3	1	
Black-cheeked Waxbill (Black-faced Waxbill) *Estr. erythronotos* 112			A										3	1	
Swee Waxbill *Estrilda melanotis* 20,62,65,68-9,70,81,121	N	N	A	1		3			1	2				1	
Quail Finch (African Quail-Finch) *Ortygospiza atricollis* 68			A	1	1		2	1				2	3	1	
Red-headed Finch *Amadina erythrocephala* 110	N		A	1	1						2	3	1	1	2
Pin-tailed Whydah *Vidua macroura* 44,64			A	1	1	2		1	2	2	2	2	1	1	1
Shaft-tailed Whydah *Vidua regia* 110,113	N		A										1	1	
Paradise Whydah (Eastern Paradise-Whydah) *Vidua paradisaea*			A	1										3	
Steel-blue Widowfinch (Village Indigobird) *Vidua chalybeata*			A		1									3	
Chaffinch *Fringilla coelebs* INTRODUCED 20,31			A	1		1								3	
Yellow-eyed Canary *Serinus mozambicus*			A	1	1										
Black-throated Canary *Serinus atrogularis* 94,101,114			A	1	1	1	1	1	1	1	1	1	1	3	
Cape Canary *Serinus canicollis* 23,46,56			A	1	1	1		1						1	
Forest Canary *Serinus scotops* 8,17,19,69,70,121	E	E	A	1		1								1	
Cape Siskin *Pseudochloroptila totta* 6,14,19,20-4,33,61,65,69,70,81,123	E	E	A	1		3		3	1	3				1	
Black-headed Canary *Serinus alario* 7,13,45,74,77,86,89,94,97,103-5	E		A	1		3	3	1	1	1				1	
Bully Canary *Serinus sulphuratus* 18,25,56,62			A	1	1	1		1		1	1	1	1	1	
Yellow Canary *Serinus flaviventris* 44,47,50,52,64,77	N		A	1	1	3		1	1	1	1	1	1	1	2
White-throated Canary *Serin. albogularis* 47,50,54,56,76-7,81,98,103	N		A	1	1	3		1	1	1	1	1	1	1	1
Protea Canary *Serinus leucopterus* 6,41,56-7,62,74,81-4,88,117,123	E	E	A	1		1		2	2	3				1	
Streaky-headed Canary *Serinus gularis* 56,69,83			A	1		2		1						1	
Golden-breasted Bunting *Emberiza flaviventris* 113			A										3	3	
Cape Bunting *Emberiza capensis* 23,47,50,81,89,123	N		A	1	1	1		1	1	1	1	1	3	2	
Rock Bunting (Cinnamon-breasted Bunting) *Emberiza tahapisi*			A	1		3							3	3	
Lark-like Bunting *Emberiza impetuani* 77-8,91,95,110	N		A	1	1			2	1	1	1	1	1	1	

RECOMMENDED FIELD GUIDES AND REFERENCES

Birds:

Sasol Birds of Southern Africa. (Second edition, 1997). Sinclair, I., Hockey, P.A.R. & Tarboton, W.R. 1993. Struik Publishers, Cape Town. Also available as *Illustrated Guide to the Birds of Southern Africa*, New Holland Publishers. *The recommended field guide.*

Newman's Birds of Southern Africa. Newman, K.B. 1983. Southern Book Publishers, Johannesburg; new edition, 2000, Struik Publishers, Cape Town. *An excellent field guide.*

Roberts' Birds of Southern Africa. Maclean, G.L. 1993. John Voelcker Bird Book Fund. *The classic handbook for further information on birds of the region; thorough new edition currently in preparation by the Percy FitzPatrick Institute of African Ornithology.*

Collins Illustrated Checklist: Birds of Southern Africa. Van Perlo, B. 1999. HarperCollins, London. *Broad geographical coverage, but pictures small and text brief.*

Bird call tapes:

Southern African Bird Sounds. Gibbon, G. 1991. *Set of six tapes, comprehensive and highly recommended.*

Rare Bird Calls of Western South Africa. Cohen, C. 2000. *A selection of calls of localized endemics, including subspecies, not available elsewhere. Contact **Birding Africa** for details.*

Site guides:

Birds of the South-western Cape and Where to Watch Them. Peterson, W. & Tripp, M. 1995. South African Ornithological Society and Cape Bird Club, Cape Town. *A thorough guide to major and minor birding spots within 300 km of Cape Town.*

Where to Watch Birds in Southern Africa. Berruti, A. & Sinclair, J.C. 1983. Struik Publishers, Cape Town. *An excellent book, but nowadays scarce and outdated in many respects.*

Top Birding Spots in Southern Africa. Chittenden, H., (ed.), 1992. Southern Book Publishers, Halfway House. *Coverage good and focus is on lists, rather than detailed site treatment of how to find birds.*

The Atlas of Southern African Birds. Edited by Harrison, J.A., Allan, D.G., Underhill, L.G., Herremans, M., Tree, A.J., Parker, V. and Brown, C.J., 1997. BirdLife South Africa, Johannesburg. *A superb handbook featuring excellent distribution maps and text.*

Atlas of the Birds of the South-western Cape. Hockey, P.A.R., Underhill, L., Neatherway, M., Ryan, P., 1989. Cape Bird Club, Cape Town. *A wonderful reference, nowadays scarce.*

Mammals:

Field Guide to the Mammals of Southern Africa. Stuart, C. & Stuart, T. 1998 (second edition). Struik Publishers, Cape Town. *Excellent field guide, best for the region.*

The Kingdon Field Guide to African Mammals. Kingdon, J. 1997. Academic Press, London. *Excellent guide to African mammals.*

Reptiles and Amphibians:

Field Guide to Snakes and Other Reptiles of Southern Africa. Branch, B. 1988 (Third edition, 1998) Struik Publishers, Cape Town. *The only comprehensive field guide.*

South African Frogs – A Complete Guide. Passmore, N.I. & Carruthers, V.C. 1995 (Revised edition). Southern Book Publishers and Witwatersrand University Press, Johannesburg. *The only comprehensive field guide.*

FURTHER READING

Liversidge, R. 1996. **A new species of pipit in southern Africa.** *Bulletin of the British Ornithologists' Club* 116(4): 211–215

Ryan, P.G. 1996. **Barlow's Lark: a new endemic lark for southern Africa.** *Africa: Birds & Birding* 1(4): 65–70.

Ryan, P.G. & Bloomer, P. 1999. **The Long-billed Lark: a species complex in southwestern Africa.** *Auk* 116:194–208.

PHOTOGRAPHIC CREDITS

Abbreviations: CBC = Cape Bird Club; Gallo Images: GI; NDWP = Nigel Dennis Wildlife Photography; PAPL = Photo Access Photographic Library; RDPL/AI = Roger de la harpe/AFRICA IMAGERY; SIL = Struik Image Library. Photographic credits on each page read from top to bottom, and from left to right.

Shaen Adey/SIL: 22, 29b, 41; Mike Allwood-Coppin/NDWP: 38/39, 108b; Gerry Broekhuysen /CBC: 26, 32c, 33c; J.J. Brooks/PAPL: 57a, 125a; Tony Camacho: front cover, bottom; 20, 64c, 91a, 116c; Cape Bird Club: 21b; 48; Callan Cohen: 8a&b, 11, 17a, 25, 34, 45, 46, 50, 53–56, 57b, 63b, 64a, 87, 88, 90, 92, 93a, 94b, 95a, 99b, 100b, 101, 102a, 103, 105b, 111a&b, 113, 115, 122b, 123, 124a, 136; Gerhard Dreyer/SIL: 47; Bruce Dyer: 37a, 40c, 104; RDPL/AI: 18b, 66b, 118; Roger de la Harpe/SIL: 18a; Nigel Dennis/NDWP: 32a, 33a, 59a, 65, 76, 102b, 108a, 110a, 116b, 119a&b, 121, 124b; Nigel Dennis/G I: front cover, top; Nigel Dennis/SIL: 109, 110b&cc; Richard du Toit: full title page; 116a; Albert Froneman: 13a, 32b, 83b; Clem Haagner/GI: 105a; John Harvey/CBC: 75b; M. Kahl/FitzPatrick Inst.: 40a; Walter Knirr/PAPL: 98; Walter Knirr/SIL: 4; Johann Knobel: 96a; Derek Longrigg: 23, 57d, 85c; Rita Meyer: 13b, 15b, 84, 117; Nico Myburgh: 7b, 17b, 42, 69, 73c, 81a, 85a, 86, 89, 120, 125b; Colin Paterson-Jones: 6, 43, 63a, 112; Anton Pauw: 33b; Peter Pickford/SIL: 29a; Neil Preyer: 79b; M. Reichardt: 116d; Phillip Richardson/GI: 107; Peter Ryan: 21a, 40b, 40d; SIL: 59b; Mark Skinner/PAPL: 30; Claire Spottiswoode: 7a, 9, 10, 24, 28, 36, 37b, 49, 60a, 66a, 67b, 68a, 71, 75a, 77, 79a, 80, 81b, 91b, 95b, 136a; David Steele/PAPL: 62; Peter Steyn: 52, 57c, 60b, 72a, 73b, 78a&b, 94a, 96b&c, 99a, 100a, 105c, 122a; Warwick Tarboton: 72b; C. J. Uys/CBC: 64b, 67a, 73a, 85b, 93b; Hein von Hörsten/SIL: 15a, 83a; Patrick Wagner/PAPL: 68b.

ABOUT THE AUTHORS

Callan Cohen and Claire Spottiswoode are both research students in evolutionary biology at the Percy FitzPatrick Institute of African Ornithology at the University of Cape Town, specialising in avian systematics and behavioural ecology respectively. They are acknowledged to be among the Cape's leading birders, have led bird tours in the region for leading international tour companies, and currently hold the record for the most bird species seen in western South Africa in one day.

Both are experienced southern African birders, and were successively the youngest

Callan Cohen

people to have seen over 800 species in this region. Callan and Claire have also birded extensively in little-known parts of the African continent, a highlight being their involvement in the rediscovery of *Namuli apalis* in northern Mozambique, not seen since its discovery in 1934.

They have contributed to many ornithological publications, including numerous books and popular and scientific journals. Callan is Vice-Chairman of the Cape Bird Club, Africa's largest bird club. Together they run Birding Africa, a Cape Town-based bird guiding and consulting company.

Claire Spottiswoode and Thyolo alethe

USEFUL CONTACTS

International callers: when dialling a South African number replace (021) with +27 21 and 083 with +27 83.

Birding Africa (up-to-date information, advice and tours); Callan Cohen & Claire Spottiswoode, 21 Newlands Road, Claremont, 7708, Cape Town; Tel: (021) 683-1898; Mobile phone: 083-2560491; Fax: (021) 671-2990; Internet: www.birding-africa.com; info@birding-africa.com
BirdLife South Africa; P.O. Box 515, Randburg, 2125; Tel: (011) 789-1122; Internet: www.birdlife.org.za
Cape Bird Club (branch of BirdLife South Africa); P.O. Box 5022, Cape Town, 8000; Tel: (021) 558-7381; Internet: www.capebirdclub.org; info@capebirdclub.org
Flowerline (MTN); Tel: 083-9101028
Northern Cape Nature Conservation Service;
Private Bag X6102, Kimberley, 8300; Tel: (053) 832-2143.
Percy FitzPatrick Institute of African Ornithology, University of Cape Town, Rondebosch, 7701; Tel: (021) 650 3290, Fax: (021) 650 3295; www.uct.ac.za/depts/fitzpatrick.

SAFRING, Avian Demography Unit, University of Cape Town, Rondebosch 7701; Tel: (021) 650 2422; Internet: www.uct.ac.za/depts/stats/adu/; safring@maths.uct.ac.za
South African National Parks; P.O. Box 7400, Roggebaai, 8012; Tel: (021) 422-2810/6; Fax: (021) 424-6211; Internet: www.parks-sa.co.za; reservations@parks-sa.co.za
Western Cape Nature Conservation Board; Tel: (021) 945-4570; Fax: (021) 945-3456; svinfo@cncjnk.wcape.gov.za
Whale hotline (MTN); Tel: 0800 22 82 22.
Tourism Offices: Cape Town Tourism; Pinnacle Building, corner of Burg & Castle streets, Cape Town, 8001; Tel: (021) 426-4260; Internet: captour@iafrica.com
Other tourism offices: West Coast, Tel: (022) 433-2380; Overberg, Tel: (028) 214-1466; Namaqualand, Tel: (027) 712-2011, Northern Cape, Tel: (053) 832-2657, Garden Route, Tel: (044) 873-6314.